LOCAL TELEVISION

Finding a Voice

by
Roger Wilson

with an Introduction by
Jon Davey
Director of Cable and Satellite
Independent Television Commission

Local Television
Finding a Voice

First Edition July 1994 ISBN 1873601 077

Published by Dragonflair Publishing
PO Box 5, Church Stretton, Shropshire, SY6 6ZZ,
United Kingdom.

CONTENTS

FOREWORD

This book has taken nearly two years to complete from the first idea, although the work from which it has been drawn has taken ten years to gather together in one place.

My first involvement in cable came in 1984 with the invitation to join the consortium which formed Cable Camden, later to become Cable London. We researched and wrote four successful franchise applications and when I left Cable London in 1992 the company was a leading operator in the UK industry.

For cable enthusiasts the journey to 1994 has been a tough one. For those believing in local programming it has possibly been the most deeply rutted course of all. Whatever the problems the industry has faced (and they have been many) the fact is that civil engineering and local television have little in common - despite the shared label of 'cable'. The commitment, experience and skills required to build high quality telecommunications networks do not usually sit easily with the more ephemeral world of creative TV programme-making.

Times are changing. There are still many hurdles to be overcome, but we are already seeing growth in local services as the need to offer variety, and to exploit the uniqueness of cable, comes more to the fore. Some of the lessons of the past have been quite painful. If this book helps others avoid some of the traps this book will have been an exercise well worth the effort.

I believe local television is a major theme in communications which has yet to be discovered. It has the opportunity to become a 'third force' in UK television - all the more important as the leading television companies seek to make their mark on a global stage. I hope some of my enthusiasm comes over in the pages which follow.

It remains to thank all those people who have helped with my research and contributed to this book. Their part in detailing their experiences has been both valuable and enjoyable for me. There is a book of anecdotes in the history of local programming if anyone cared to write it !

Their enthusiasm has been reflected by those who have helped more closely in the development of the book. I should like particularly to thank Rob Bason of Learning Television. He has been the perfect foil for all sorts of ideas, drawing on his varied experiences with Coventry and Milton Keynes. My grateful thanks also go to Joyce Taylor (of United Artists) and Shan Stephens, who have been kind enough to allow me to quote extensively from their own original writing. Finally I am grateful to Julia Duckworth, who has been a critical reader bringing her special experience and eye to the text, and to Chris White, for his creative cover design.

Most importantly I must thank Jon Davey of the ITC. Without his encouragement and self-effacing advice, this book would not have been possible and could not have been as comprehensive as I hope readers will find it.

Roger Wilson
June 1st 1994

INTRODUCTION

by
Jon Davey
Director of Cable and Satellite
Independent Television Commission
(formerly Director General of The Cable Authority)

For the Annan Committee in 1977, the future of cable television lay with local community services. Unfortunately for that vision, however, the Committee's Report coupled it with a reluctance to allow cable access to commercial services which would underpin a service of community television and provide a real justification for the building of new systems.

In the Committee's perception there was a strategic choice between pay television and community television as the way forward for cable systems. Since the Report did not hesitate in rejecting the idea of pay television - in the Committee's notable phrase it was a "ravenous parasite" - the benefits of community services were promoted by Annan but without any discussion as to how they, or the means by which they would be distributed, would be financed.

By the time the future cable policy came to be considered again, in 1982, it was plain that a more pragmatic approach was needed. This time new objectives other than community television had been identified - the information technology benefits that cable could bring - but it was clear to all that pay television would be a crucial part of the reason for investing in cable and subscribing to it. Without creating the circumstances in which cable could become a successful business, there was little point in discussing how it could serve community interests.

That thinking underlay the drafting of the Cable and Broadcasting Act 1984 and the franchising policies and decisions of the Cable Authority. It continues to influence the approach which the ITC takes towards cable today.

Moreover, for most of the intervening years, there was considerable doubt, certainly among potential investors, as to whether cable would be a successful business in the United Kingdom. Few franchises were able to gain the financial backing they needed. None of the early broadband operations were demonstrating the kind of performance likely to convince sceptics that investment would reap a bountiful return.

These were the circumstances in which it would not have been sensible for legislators or regulators to seek to insist on services which were uncommercial or, even worse, which were all too likely in the early stages of a cable operation to involve costs calculated to destroy the viability of the business and to bring it down. Hence the Cable Authority's ready acceptance, as recounted by Roger Wilson in his valuable survey, of Aberdeen Cable's decision in October 1985, when it was struggling to reach its first thousand subscribers, to close down its over-ambitious local programme service.

So we had a statute that imposed no requirements for mandated services but merely required whatever proposals were made to be taken into account in franchise decisions. The regulatory body looked at proposals largely from the standpoint of commercial realism, applauding an interest in and commitment to local television but more often questioning applicants about how they could justify their proposals rather than trying to squeeze more out of them. Sometimes, it has to be said, they awarded a franchise in the belief that the applicant had over-estimated the kind of local services which he could realistically support, discounting any possibility of keeping him to them.

Much of what has actually been achieved over the last few years has been impressive and encouraging, even if there have been disappointments and frustrations. There have been local television programmes in which to take pride, exhibiting imagination and skills, commitment and quality in a way that confounds the arrogant assertions that used frequently to be aired that watchable television demands production values that only terrestrial broadcasters could afford.

That there have not been more programmes is hardly surprising, given that cable subscribership, even across the whole of the United Kingdom, reached half a million homes only in the second half of 1993. The imperative for almost all cable operators has been to concentrate on getting a system built and, in all too many cases in the past, to find sources of finance with which to do it. Devoting resources to establishing local television services has not been a widespread priority.

That is changing. The cable industry is out of the doldrums. Reinforced by confidence that the combination of two businesses, cable television and telecommunications - not to mention what may be developed in the fertile ground where they merge - make UK cable an attractive prospect, investment is increasing. The rate of construction is accelerating, audiences are growing and the industry is able with greater assurance to raise its eyes from short-term crises to some of the longer term issues which will govern the success of cable.

Local television is one of those. It is a service which cable systems can uniquely provide and which some operators have already found can generate an astonishing response. While it is not easy to quantify its commercial value, it represents a significant way of differentiating a cable service from rivals, and of developing both community interest in what cable offers and customer loyalty. And the maturing of the industry enables regulators to begin to look more closely at the prospects for implementing those franchise aspirations so commonly expressed.

Roger Wilson's study is timely. Apart from considering the issues in depth, he has reviewed what has already been done, and how, and has included pointers to the ways in which a worthwhile service can be instituted. With increasing numbers of cable operators set to explore the provision of local television services, the comprehensiveness of this review of the background and of the experience so far will be of inestimable value and assistance. It should help to light the path, to prevent others' past mistakes from being repeated, to offer a source of ideas and advice and, it is also to be hoped, to provide some inspiration to develop local television services.

CHAPTER 1

JUST NICE ?

Why is local programming important to cable ?

By their very nature cable systems are able to offer local services. They provide dedicated connections into individual homes within a specific area and, whatever the technology purpose of the cabling, these lines all cluster into a local main technical centre. This is the nature of the telephone network and of the earliest cabled entertainment systems which were developed in the early years of the 20th century to carry sound-only services in major city centres. The growth of radio as we understand it today came with 'wireless' transmission - broadcasting - offering a universal service.

By contrast the later 20th century development of TV is following an opposite path. It started with broadcasting in the 1930s and the need to offer universal reach which in turn would justify TV set manufacturing. But the airwaves have limited capacity and endless demands upon them. By the end of the century cable TV systems in the UK will be serving the needs of 20% of all households, and in Europe and North America an even greater proportion. Delivered in this way television has the opportunity to acquire the local context which since the 1970s in radio has fuelled expansion and innovation, and which terrestrial television broadcasting has never quite been able to manage.

Local television is only one element in that TV revolution and one which it is hard to grasp. In turn the TV revolution is only one part of the new telecommunications ecology which is evolving in the later 1990s. The powerbase of communications is changing because of a convergence of technologies leading, ultimately, to integration based on digital techniques. A range and variety of new opportunities are opening up, some are already being exploited and some will only be realised when new concepts of communication are developed. All these opportunities are challenging, but few have the combination of immediate audience understanding, attraction, the potential for contention or, it must be said, the difficult economics that local television services present.

Who Can Benefit from Local TV

A technology based opportunity is not on its own a reason for developing a new service. Someone must see benefits and someone must invest time and finance.

Commentators immediately see that education can benefit from the changes in the local communications infrastructure. They do not always identify how structures will evolve to meet its costs and management demands. Local libraries and other public information providers (Councils, training agencies, welfare services etc.) can also be beneficiaries. So too can Chambers of Commerce and major local service providers (transport, utilities etc.) - indeed anyone with large amounts of information which they wish to make available to a wider audience. But none of these on its own is an argument for local television. They support the case for a new telecommunications infrastructure to deliver information (eventually accessible interactively) but do they create as convincing a case for the production and transmission of local television programming? And do they offer the local programming which audiences will want?

The Role of Cable TV Networks

Even though every successful broadband cable licensee offered local services when applying for his franchise (and was required to make statements about local services in his application for the franchise) the start up of local operations has been patchy. There is no uniform commitment and some cable companies will argue that there is no commitment at all. In many cases the franchise winner is no longer represented in the management team which runs the franchise, and the new shareholders who are funding a multi-million pound build programme do not always feel committed to fulfil promises they did not themselves make.

In many instances too, the American manager's experience of local programming from the USA is uncomfortable. Access to TV locally is defined by the courts as a constitutional right of free speech. It has sometimes been a contractual commitment from a local franchising authority. It has been used as a charter for aggravating the local audience, to the concern of the cable company. Alongside this the operator is obliged, however reluctantly, to fund it from what would otherwise be profits. Every cable operator is concerned about costs, all the time, with an unceasing concentration, so such obligations are frequently seen as an undesirable and unnecessary burden.

Even so cable operators now share one clear understanding. TV and telecommunications go together and the visionary environment which UK legislators embarked upon in the mid-1980s, recognises this integration by allowing cable companies to build the infrastructure to carry both kinds of service, and to compete with previously well-established service providers.

The technologies are coming together, first in the physical mechanism of delivery (single cable subscriber drops for telephone and TV are already made) and, as the 1990s progress, there will be true integration using digital technology. Whether the audience will understand all this is unlikely, that they will be concerned about it is improbable, that they could gain from it is undoubted.

Television entertainment and information will always be high on the list of domestic priorities but the means by which it arrives on the screen will be a matter for individual household judgement based on cost, variety, ease of access and quality - not of technology choice. Similarly the telephone will continue to be a key item of household expenditure - but even if an integrated service provider delivers both TV and telephone they will be seen as entirely separate services. How correlation between the two will develop in the subscriber's understanding as 'interactive television' develops is a matter of conjecture.

Reaching for Convergence

In the same way that local telecommunications services are growing in importance in the local economy, reflecting their importance in the world's economy, so local television services will become an important part of the future of television information, education and entertainment - itself increasingly a global business.

If, at the local level the two kinds of service become technologically indivisible, how should a cable operator distinguish between TV and telecommunications services?

Why should he fund local TV services any more than he funds a telephone conversation - they are both just users of the bandwidth he is selling?

He can view his role simplistically. He can provide the mechanism which delivers the service capacity, and leave others to develop or provide the service itself. It would be a valid approach.

Funding Cable

The cash demands of a typical cable franchise are heavily loaded towards the start of the operation. One estimate suggests that an average sized franchise (100,000 homes) costs between £2-3 million before any customer cash comes into the business and it will be several years and many millions later before the operation moves into an operating profit. Once operating profitably investors can attract loans to help fund the rest of the build. The total investment involved will be more than £40 million, and net cash outflow does not stop until the build is completed. Only when loans are repaid can shareholders hope to see a return for their early efforts. The returns then could be very satisfying.

The news that the first billion pounds invested in the UK cable industry was spent getting 430,000 TV customers and connecting 30,000 telephone lines (or some £2170 per connection) does beg the question of how much impact the next billion will have.

This position will be reached by the end of 1994 and there will be 1-1.2 million customer connections by then, of which probably 40% will be telephone lines. These will represent something more than 2% of UK TV households and about 1.5% of all telephone line connections. The investment per connection will be dropping towards £1500, a significant decrease, but still a very costly business. When the current franchises were awarded the average cost per subscriber connected in a fully built franchise would only have dropped below £1400 when more than 30% of homes passed by the system actually subscribed. The importance of the role telephony has taken in giving reasons for funding UK cable is also emphasised when the 'churn' (ie. subscribers signing off) is taken into account - it is far lower for telephone services than it is for cable TV.

Satellite Television

This is where a comparison with the experience of direct-to-home satellite TV, often seen as the competitor to cable television, becomes valid.

Satellite TV was oversold to its early investors, all of whom were expecting quick profits, and were thus prepared to invest (and risk) substantial sums. Profits are now coming as services grow, subscribers are prepared to pay more, and they create an attractive advertising market - about 5 years on. Costs have been rigidly controlled and every opportunity to expand quantity (and thus saleable airtime) at minimal pro-rata cost has been taken.

By contrast the potential of cable was understated until 1993. Even so the short term demands of capital markets do not easily match the shape of the cable business. Conventional investment wisdom says do not own expensive assets if they do not earn appropriate returns quickly, and do not provide loans to build such assets until the operating company can demonstrate that it can pay the loan interest out of revenue. Funding system development is thus left to the cable operator's own resources until his operating cashflow is positive.

The Cabinet's Information Technology Advisory Panel (ITAP), reporting in 1982, recommended that cable should be privately funded. Although ITAP understood the scales of investment that would be required it seriously overestimated the willingness of UK investors (corporate or institutional) to fund something of this kind. The March 1984 removal of capital allowances by the Chancellor of the Exchequer undermined this fragile position. The Channel Tunnel was a more tangible project requiring a similar scale of investment and over a similar timescale but the technology context was more familiar. Funding that project proved difficult enough for a single company and the inevitable construction delays created constant tension between the company, its contractors and its backers.

Cable versus Satellite

Viewed this way the comparison between cable and satellite is quite revealing. Satellites also cost large sums but the system is available to its entire potential audience once the satellite has been launched. But it is a one way technology with a finite life. The emphasis therefore is on supplying programming which encourages as many people as possible to buy the necessary dish.

Provided that cable can supply the same programming the competition that the press perceives between cable and satellite is largely an illusion. Where there is direct competition cable companies have tended to win because they offer additional services satellite cannot provide. Satellite systems can only guarantee programmers and advertisers an audience, and only guarantee an audience a wider range of choice from globally distributed services.

Cable systems provide a range of services to their customers of which global programme provision is just one, voice telephony is another, and local services a third. The cable operator is a local retailer selling to large numbers of individual subscribers, and some of his services are exclusive. A satellite TV provider is a direct marketing company with a more limited offering operating in a national market. Thus we find that the cable company has two separate and distinct roles, and is a very different company from what was envisaged in the early 1980s.

The first role is the development and provision of an infrastructure - an exclusive right which also carries obligations under the terms of the operating licences. The second role is to gain customers for this infrastructure. He can do this in two ways - as a distributor/reseller of other peoples services, and as a developer/provider of services he owns himself. The main range of retail services which attract his initial customers comprises satellite TV programming and telephony where he competes with BT. His search for a distinctive 'edge' over his competition entails taking 'cable exclusive' TV channels (channels which are tape delivered, are debarred from direct satellite subscribers through encryption or are carried on non-accessible satellite transponders), and in competing with BT on price and service quality. A few have also moved into the provision of local television.

The Rewards

Local television is a cable exclusive service. We have seen some of the community benefits that local services can create but given the above analysis of a cable operator's role, where are the rewards for him in local television? Where are the organisational models that help him manage its start-up and operation and, above all, where is the evidence of financial viability for local TV operations?

The UK's local programming story started in the 70s, moved up a gear in the mid-80s, nearly ground to a halt in 1989 and has grown slowly but steadily since with alternating cycles of encouraging and discouraging news. There has been little real growth, there have been a few successes, but some failures and withdrawals. There has been a lot of learning, but no meaningful large scale introduction despite the promises of franchise applicants.

The above financial arguments illustrate why this is so. The demanding financial background to a cable operation, the relative immaturity of an industry which is not yet reaching a significant audience, and managements stretched by significant operating demands, do not provide a good base for developing innovative TV services of any kind. Nor do they provide a good base for developing any innovative services. While many senior managers in cable believe that they understand TV, the issues of developing a new television channel with a range of original programming are very different from those of building networks, selling to subscribers, or developing products in any other commercial enterprise.

Cable operators are fully stretched coping with the details of daily needs. The build involves them in civil engineering, network technology design and installation, rigid quality control, and maintaining on-going local relationships seriously affected by the fact that they are digging up all the local roads. Marketing and selling involves them in activities as diverse as business telecommunications, TV product acquisition (not made any simpler by European copyright legislation), recruiting, training and managing door-to-door and tele-sales teams. On top of this is the need for customer installations and maintenance, billing (and collecting) small accounts from large numbers of people, and the huge demands of setting up and running an efficient customer service operation. Quite enough for anyone. TV programme production, channel management and advertising sales, even in a relatively loosely regulated structure, are yet more burdens. If they cannot be financially justified the conventional wisdom is - don't do it.

There is also the belief that TV programme production in the UK is expensive. This was certainly once true but it has been eroded with the growth of the independent sector, wider cost consciousness from national and international TV channels, and the reduction of union power through a combination of technology evolution and legislative action. Nonetheless the high cost perception remains and reinforces the conventional wisdom.

There is some evidence to show that the rewards from local programming are real. As networks grow operators search for a wider range of services. Customer 'churn' is a significant performance problem for all cable TV networks, and building customer relationships to reduce the 'come-and-go' attitude is a critical need. Most operators believe that telephony helps and there is evidence to support that view. Local programming should be more actively considered as another retention marketing opportunity.

Research has shown how attractive local TV is to the audience in the theory. Is the practice any different? It is often acknowledged that local TV as a customer retention 'weapon' is likely to be more promising than as an initial sales tool, particularly in the shorter term.

Local TV as a revenue generator is dependent on a sizable enough audience to make the proposition attractive to advertisers. Establishing it as a local advertising medium takes time - so starting sooner rather than later brings benefits quicker. Local TV as a medium for promoting the local identity and public responsibility of the cable company is a less quantifiable measure - but may be equally valid.

The added value to cable systems that different methods of local channel organisation are beginning to demonstrate stems from innovative managers, defying UK traditions of TV production and management. Some of these approaches come out of the whole picture of UK cable TV experience - now also coloured by the influence of Canadian local TV, with probably the simplest structure and most experience of the medium in the world.

The uniqueness of the UK regulatory structure and the regional concentration of system ownership which has evolved in the UK cable industry provide other important influences.

This is our context - reaching out beyond experiments to the stage where local television channels, offering value to communities, programmers, advertisers, channel operators and cable companies can operate as commercially viable, socially valuable and customer-valued components of the total TV environment. And all this at a time when technology is providing the foundation for huge changes in the global telecommunications environment - changes which may place an increasing emphasis on local identity, local delivery and local service.

CHAPTER 2

BUILDING A CABLE INDUSTRY

The experiences of the 1970s and the early 1980s

Before looking at the recent local programming efforts of the late 1980s and 1990s it is will be valuable to recall the nature of the UK's television industry and consider some of the aspects of cable TV operations and programming during each stage during the growth of cable operations.

British TV had a single channel from the British Broadcasting Corporation (BBC) until 1955 when Independent Television (ITV) began transmission, initially in London and Manchester and slowly spreading across the country to 100% reach. BBC2 began transmission in 1964, using a new set of technical standards (625 line UHF) and moved into colour from 1967 onwards. In 1969 both BBC1 and ITV began 625 line and colour, with transmissions in parallel with the older 405 line, black and white format. The new TVs necessary for colour were in the shops and the plan was to phase out 405 line VHF as soon as possible (it eventually ceased in the mid-1980s). Channel 4 arrived in 1983.

The growth of television in the 1950s was necessarily slow and in order to allow as many homes as possible on the edge of transmitter areas to access the new service as quickly as possible the Home Office (the regulator) licensed cable systems, which were commercially funded. Many local authorities also built cable systems so that housing blocks did not need to be festooned with aerials. Thus was born a cable industry. It received a revitalising boost with the introduction of colour, although it was acknowledged politically that its long term existence was likely to be tenuous as both the BBC and the ITA (Independent Television Authority) planned nationwide coverage with their transmitter networks. It was this finite life that led to the first stage in the life of local television. In 1972, after intense lobbying by the Cable Television Association (CTA), the Government permitted five of the larger cable networks to take licences for a local television channel. The objective was to see whether a new form of television could help to insure the future for cable networks facing the growing penetration of the terrestrial transmitter network.

Maurice Townsend had been working in cable in Canada (he is an accountant as those involved today with the CTA over the years have been gratefully aware). He came back to the UK in 1970 joining Greenwich Cablevision to work in marketing and management, and then to be its Chief Executive. When in 1972 the idea of running a local television channel to create a competitive edge for cable came up he became an active member of the lobby.

The five licensees were Sheffield, Wellingborough, Swindon, Bristol and Greenwich. Greenwich Cablevision (which actually served the less attractively named Plumstead area) was the first (in 1972) to receive its licence, which included the right to produce and show local programming, and it was the first 'on air'.

Perhaps it was inevitable that Maurice became an unofficial public spokesman for the local television industry as well as a prominent CTA advocate. The Greenwich licence covered some 20,000 households south of the River Thames on the south-eastern edge of London and was the only large cable system, and the only of the five multi-channel relay systems, within easy reach of central London. His campaigning alongside a local MP with front bench experience, Christopher Mayhew (now Lord Mayhew) gave him access to Parliamentary influence, and he is also always good for a quote. This flag waving made him friends and enemies, the former largely local and the latter mainly national, not least in the broadcasting establishment which could never quite work out what Greenwich Cablevision was about.

Some of those running other early local channels are now as well known in their own way. Richard Dunn (who managed the Swindon local channel) has achieved a pinnacle in TV, as managing director of Thames TV and as a former chairman of Independent Television News. When his experience of local channels is remembered Thames' commitment to the City TV concept for the Channel 5 franchise may come as no surprise. It brings him full circle.

The five local channel licensees had a particular brief. They were to provide TV channels to a specific area where newly introduced broadcast colour TV was poorly received. These local channels would be the fourth channels in their area and would be subject to all the same regulations as ITV, but their regulator was the Home Office, not the professional regulators of the ITA. There was no intention of allowing broadcaster freedoms to these local channels though they were free to acquire programming in the international market, and to make their own programming in just the same way as the ITV companies.

A few hours a week of local interest programming was impossible to produce at a low overall cost and to a high technical quality during the 1970s. Even low-grade camera equipment was cumbersome, expensive and required expert hands-on attention in operation. Video editing was crude (much broadcast TV still used film when sophisticated post-production was required) and the TV industry was unionised and rife with restrictive practices, tacitly supported by employers in ITV for whom such costs were allowable against both tax and the Treasury levy on commercial TV profits. Despite this there was much innovation and these local channels were certainly viewed. Advertisements were permitted and Greenwich at one stage recovered its cash costs, while providing a useful stimulus to the local economy.

The sad fact is that despite the innovation and enthusiasm this first stage in local television failed for one very good reason - cable cost subscribers a rental, putting up an aerial was a once-for-all expense. Without being able to point to 'added value' the cable companies could not win, and local television was not perceived as sufficient added value by subscribers. A local channel could not reach enough subscribers to generate enough revenue to make or acquire the programming which could compete with the broadcasters on variety or quality. Costs could not be cut because of the underlying capital costs and the restrictive practices enforced by the unions. The Bristol and Wellingborough channels were closed in 1975, Sheffield followed in 1976. A sixth local channel (Channel 40) was licensed in Milton Keynes in the same year but it too closed in 1979.

The TV channel competition was growing fiercer all the time. By early 1981 transmitter coverage in the cabled areas was virtually 100% and a fourth broadcast channel was being planned for a launch in 1983.

The Swindon local channel closed in 1981 though memories lasted and its resurrection was locally welcomed. The last of these first local operations to close was Greenwich in 1984 - still run by Maurice Townsend, though he could see the end looming for the entire Greenwich network, under pressure from terrestrial broadcasting and the new legislation which would create broadband cable franchises. John Marriott (now film critic for Sky Television and for local radio in London) joined Greenwich Cablevision for a short period in 1983 and remembers that there was a sense of mission and a strong belief in the future of cable throughout the company. Most of the manpower now came from volunteers (under the auspices of the Greenwich Television Society). Regular programmes then included a news magazine three times a week, coverage of council meetings, and local documentaries. The 3-camera studio and two location units worked with low-band U-matic equipment and all the station's staff had some technical or production expertise.

The cable companies faced a situation where they had no marketing edge. Political pressure again resulted in new licences, this time to run locally scheduled subscription channels offering movies and entertainment programming. Thirteen of these licences were awarded in the autumn of 1981, including channels in all of the previous local areas. Amongst the newcomers programming these channels was a Milton Keynes company, SelecTV, which in the 1990s has grown to become one of the the largest independent producers and whose name crops up again later in the tale of cable TV programming with Channel One in London.

Many names still active in the industry became prominent during this time - Richard Wolfe (later of The Children's Channel and now with the Travel channel) created the Thorn EMI subscription channel in Swindon and Nick Mellersh (an independent consultant) was responsible for Rediffusion's programming. The Home Office official responsible for handling the licensing of these pay TV channels was Jon Davey.

But the future of cable was now climbing the national agenda. Fierce lobbying throughout the 1970s had failed to save the relay industry and now broadband technology was looming. In the USA broadband was accounting for phenomenal investment in cable systems carrying 20 or 30 channels - an unbelievable number in those days. The visions went wider than TV. These were the days when the gurus who advise governments saw information technology providing a new industrial revolution.

Unfortunately for cable TV the process the UK government entered into finally put cable developments lower down the list of national priorities (and therefore for parliamentary time) than their own gurus recommended. When the Cable and Broadcasting Act of 1984 finally arrived on the statute book and Jon Davey took up his post as the Cable Authority's Director-General three years had slipped away.

The ball had been rapidly pushed to the top of the hill by the Information Technology Advisory Panel (ITAP), set up to advise the Cabinet in late 1981. ITAP was made up from five leading information technology industrialists. To the surprise of many it took cable as the subject of its first report and captured the political imagination. Despite many public misgivings about the potential impact on the public service commitment in British broadcasting, matters started to move ahead with some urgency.

It is worth recalling part of the statement of introduction that ITAP made in its report to the Prime Minister of the time, Mrs Thatcher. (HMSO: February 1982: ISBN 0116308214):

"We are convinced that there are powerful economic and industrial arguments for encouraging cable systems in the United Kingdom. Further, we believe that, given the right conditions, these could be entirely financed from private sources. Our report sets out our reasoning and makes specific recommendations for steps which we think should be taken by Government in order to implement such a policy in the fastest possible timescale. We emphasise the need for urgent decision and action. The opportunities which such a policy offers to the United Kingdom will be irretrievably lost if our timescales are not met. A delayed decision will be the same as a negative decision."

The ITAP report mapped out a series of opportunities for British technology, British exports, social and economic change and assessed the danger of missing the opportunity. However tempting it may be ten years later to create a politically critical perspective based on hindsight, is not the role of this book - we shall confine comment to the prospects for local television as seen by ITAP and accepted by Government.

Local, educational and minority services were regarded as key benefits a cable system could provide. The prospect for 'direct broadcast satellite' TV (DBS) was taken very seriously and the report devotes a whole chapter to exploring the relationship between satellite and cable. Local was an attractive extra which only cable could provide and *"not only, in our view, does DBS need cable ... but DBS can provide cable with some of the extra programmes that they need."*

This reflection that cable and satellite need not be seen as competitive, but complementary in a healthy competitive market, was ahead of much of the thinking of the time.

The opportunity for local TV news was discussed, and the relationship which could result with local newspapers. The difference between news on TV and news in print was understood, but cable was clearly identified as a powerful competitor in the fight for advertising, as opposed to notions of competition in information provision.

On the broader front of community programming the report commented:

"The experiments in local programming that have been conducted in the United Kingdom have shown that a substantial interest exists in such programmes. However, those experiments could not cover their costs. It remains to be seen whether, in the context of an extended range of cable services, with correspondingly greater opportunities for subsidy of one channel by another, it will be possible to carry a full range of local news and information."

This concept of channels being able to provide each other with cross-subsidies within a cable network has never been workable, though clearly BSkyB has been able to take this approach.

Tele-shopping, interactive cable education, and advertising interests also *"clearly stand to gain from the introduction of cable systems"* though the localness of any such involvement was not expanded.

The committee acknowledged that there were a lot of issues which it had neither the time nor the expertise to tackle; not least these included the regulatory environment and implications for the broadcasters, British Telecom and the electronics industry. Inevitably the broadcasting lobby got to work with a vengeance. Cable was a threat. In addition the future of British Telecom as a state owned corporation was already under review.

The outcome was a three-man commission of enquiry appointed by Home Secretary Willie Whitelaw under the chairmanship of Lord Hunt of Tanworth to report on the broadcasting and regulatory implications of an expansion of cable along the lines recommended by ITAP. Secretary to the Hunt Inquiry was Jon Davey, then head of the Broadcasting Constitution and Licensing Division of the Home Office. One of the Inquiry's members was

Professor Jim Ring - later to be deputy chairman of The Cable Authority, previously a member of the IBA and today a member of the ITC.

The Hunt Inquiry reported in October 1982 and gave the green light to cable development, and the start of a more loosely regulated programming environment. By that time it was too late for legislation to be passed by the 1982/83 Parliament. The Inquiry report was debated in the House of Commons on December 3rd 1982 when the Home Secretary announced that a Bill would be laid before Parliament in the next session. In the meantime he accepted the need for a franchising procedure and a new regulatory authority to administer the industry and he reflected press concerns about a possible explosion of pornography . He also indicated that he would legislate to prevent exclusive acquisition of rights to national sporting events by cable companies.

Responding for the Opposition the Shadow Home Secretary, Roy Hattersley, indicated that a future Labour Government would *"want all the benefits to be enjoyed by all the people, and not exploited by a handful of foreign-based companies, who would dilute the standard and content of British television."* He called for a common national carrier - naming British Telecom - and described the Government's approach as *"undue and irresponsible haste"*.

The Government seemed to accept the need for speed. When the White Paper proposing the details of legislation was published in April 1983 it announced that, as an interim step pending the enactment of legislation the Government would invite bids for ten to twelve pilot franchises of about 100,000 homes. Bidders were able to nominate their own areas. Thirty seven applications were received, a level of interest which far exceeded expectation. Evaluating these proposals while at the same time preparing detailed legislation presented the Government with a problem. The decision was finally made to use a firm of consultants, EIU Informatics, to assess the detail and a panel of 'three wise men' to sit in final judgement. One of the three was Professor Ring.

In Autumn 1983 eleven franchises were awarded. They were the first broadband franchises:

Swindon	Northwest Glasgow	Ealing
Windsor	Belfast	South Liverpool
Aberdeen	Croydon	Guildford
Coventry	Westminster	

(Note: the Belfast franchise has now lapsed)

The giants of the old cable industry, Rediffusion and Visionhire, failed to make much impression on this initial franchise round. Of the eight franchise applications made between them, only one - Rediffusion's showcase proposal for Guildford - was successful and neither company was destined to remain long in the industry.

As the award winners laid their detailed plans and tried to raise the funding to build what they had won, the Cable and Broadcasting Bill came before Parliament. Even before it was on the statute book new ways of licensing and regulating cable television had to be put in place. The Home Office licensed the first of the existing narrow-band cable systems to carry satellite 'direct-to-home' (DTH) TV channels in January 1984 under the older Post Office Act. First was Sky Channel, followed by TEN - The Entertainment Channel, Screensport and Music Box. Many of the old 4-channel Rediffusion systems gave their customers aerials to receive the four off-air channels and offered the new satellite delivered channels over the cable. Soon the first enthusiasts satellite dishes began to go on sale, although owners needed both a reception licence and local planning permission for the dish.

The Cable and Broadcasting Act 1984 came in the second wave of new legislation following the 1983 General Election. Before it reached the statute book the search for a chairman for the new Cable Authority began and Richard Burton, former chairman of Gillette UK Ltd, accepted the task in August 1984, a few days after the legislation had received the Royal Assent on 26th July. The other members of the new body were appointed (Professor Ring was deputy chairman), they chose Jon Davey as their Director General, and in January 1985 the shadow Cable Authority came out into the light and the new regulated world of cable began.

Broadband Launches

The first eleven franchises were now getting underway. Swindon (owned by Thorn EMI) was first in September 1984, bolting on broadband expansion to the existing narrowband franchise. Aberdeen was the first new start in May 1985 and during the year Clyde, Coventry, Westminster, Windsor and Croydon followed suit.

It is worth recalling that the cable offering in those early days did not have the range and quality available today. The urge to provide local TV was fuelled by the need to build quantity and variety in the programming on offer as much as by a sense of commitment to provide local services. The range of programming was weak and the resources invested in the few satellite channels were not sufficient to allow effective competition with the mainstream broadcasters. I remember the Music Box studio - a converted conference room on the 12th floor of an office block in Tottenham Court Road. It did not even have space for an air-conditioning unit.

Thus local TV was high up the agenda of the first franchises and their experience still influences perceptions and practices in local television today.

CHAPTER 3

THE FIRST BROADBAND NETWORKS

Are you getting the best from your TV ?

The local channel was a major plank in the launch package of all the early starting broadband franchises. The main reason was that despite a 30-40 channel capacity on the network there were very few channels available. Four broadcast channels, maybe an 'out of area' ITV station, Premiere (a premium movie channel), Sky Television, Screensport, TEN The Entertainment Channel, and Music Box. They were followed later by The Children's Channel, Arts Channel, Lifestyle and SuperChannel (started by a consortium of the major ITV companies and described as "ITV without the boring bits"). There was also the opportunity to take overseas channels.

Swindon

The first of the broadband franchises to start operations was Swindon in September 1984. Swindon was an upgrade area - it had been served by cable for more than 15 years already, so although the launch was a moment in history the fanfares were reserved for the 'new build' launches in 1985.

Swindon's local channel carried news, topical programming and some bought programmes - modelled very much on the lines of the earlier pre-broadband years. The head of programming, David Burrows, is a respected broadcast producer and some exciting and innovative programmes were made. His team also claimed to be the first to take video equipment into the cockpit with the Red Arrows. This programme was shot in Lincolnshire. It is quite apparent that the geographical reach of the new local programmers was not confined to local, as long as they could find a local angle to a story.

The positioning of all of these new local channels was to compete for viewers. The cable network's priority was to develop unique strengths and a quality local output with the variety to attract viewers from the broadcast channels.

Aberdeen

It was Aberdeen's local channel which attracted all the attention. At the launch in May 1985 it featured a staff team of twenty professional editorial, production and technical staff and included leading television journalists attracted from the locally based ITV company Grampian Television.

They put together a news magazine programme every evening and covered other local stories for a local channel whose standards were as high as anything transmitted by the terrestrial broadcasters. But the subscriber numbers were tiny and with the build progressing slowly because of differences of opinion with the local authority, subscriber numbers grew at a snail's pace. It was very few weeks later that Patrick Scott, the managing director, advised Jon Davey at The Cable Authority that the cost of running a local programming service was draining the company's reserves in a way that he could not accept.

The programme making unit was hived off to become an independent production company, Tern Television. Although a local channel has survived the company's funding commitment to it has been tiny.

Coventry

At the same time John Ross-Barnard was finalising his plans for the launch of Coventry Cable and Joyce Taylor was recruited from Glasgow University to head up Clyde's local cable channel. They were both watching the Aberdeen experience with interest.

John was one of the most prominent broadcast programmers brought in to run a cable system. Former BBC Director General Sir Ian Trethowan had joined the board of Thorn EMI when he retired from the Corporation. In this non-executive role he found himself advising Thorn's video and TV division about the Coventry franchise, the first new build area Thorn was planning. Sir Ian contacted Ross Barnard who he remembered as an entrepreneurial producer, too often pushing back the commercial boundaries of the still heavily traditional, and somewhat reluctant, BBC. Some conversations later the former pirate radio presenter turned BBC executive found himself the first managing director of Coventry Cable. His brief was to create a competeitive programme package for the thousands of homes who would soon be passed by the new cable system.

Trethowan was convinced that cable TV was going to be like local radio with pictures. He may yet be proved right. Thorn EMI believed that its rental shops would sell cable TV subscriptions. They were wrong. The role of the cable station was defined as to coordinate incoming channels, produce local channels, take commercials and build an audience against competition from press and radio.

Even at the launch in September 1985, and despite the substantial investment made by Thorn EMI and British Telecom (its minority partner) the combination looked an unlikely model.

The Coventry Cable marketing strategy was to try to identify with local people at all levels. The council was seen as a key means to attract support so a way had to be found to get Council news and information on air. Coventry City's "Contact" is still transmitted today - one outstanding example of good relationships between cable and local government. The fact that a conscious decision was taken to get the council on the cable company's side clearly helped the build. Ross Barnard describes it as 'satisfying their egos' but the result is that Coventry is a city ahead of others in communicating with its local people.

The heart of start-up programming was a daily news magazine, but rather than recruiting a staff team of journalists the local radio station, Mercia Sound, agreed to co-operate. Ross Barnard believes that Coventry piloted the 'bi-media' approach which requires reporting to be as right for radio as for TV. By calling for only one set of location research, good co-ordination between radio and TV editors on the choice of story, together with efficient production techniques,two reports could be recorded on site in little more time than one. The cable station's staffing was thus technically biased rather than dominated by editorial staff.

Programme consultant Rob Bason remembers that the alliance was more impressive in theory than in practice. Experienced radio journalists often found adding a TV report to their radio work difficult to adjust to, and some of them gave little or no priority to their TV work. The TV team increasingly set off to cover their own stories in their own way.

Ross Barnard's search for a tabloid TV approach was also resisted by Mercia Sound's MD, Ian Rufus. Another key influence was the editor of the local paper The Coventry Telegraph, Geoff Elliot, who kept a close and critical eye on the channel's progress and provided a spur to make the product journalistically better. Leading TV producer Mike Hollingsworth became involved when he left TV-am and he ensured that the programme retained a common touch.

All too soon it was obvious that the ambitions were misplaced. Thorn EMI was in corporate difficulties and in a fire-sale to strip itself of development companies and project liabilities in February 1986 the Swindon system was sold to British Telecom. The local programming operation was closed and the staff made redundant prior to the sale. At the same time Thorn EMI sold its controlling interest in Coventry to its partner, again British Telecom, although the management team were able to protect programme making for a while by using the programme facility to raise revenue through training.

But it was not long before cutbacks were necessary. In 1988 the majority of the staff programme and technical team were made redundant and effort was concentrated on developing relationships with local colleges, and on volunteer based production. The channel continues on this basis today, its value being recognised by new owners Devanha.

Clyde

The broadcaster approach adopted in Coventry and Aberdeen contrasts with the plans laid by Joyce Taylor in Glasgow. The former head of television for Glasgow University won the job of local channel manager ahead of another early local TV stalwart, Tony Currie - who, failing that appointment, promptly went on to become Controller of Programmes at The Cable Authority.

One of the key investors was local radio station, Radio Clyde. Another was Robert Maxwell's Glasgow Daily Record.

Joyce visited Aberdeen and Coventry and came away believing that broadcast standard production equipment and skills were not essential for the production of quality local television. Good professional non-broadcast equipment was cheaper and offered greater flexibility. She then recruited her staff specifically from non-broadcast areas - youngsters from college, university TV engineers and those who, she said, "seemed to be able to turn their hands to anything".

The channel went on air on October 5th 1985, and just like its predecessors elsewhere in the country reached less than 100 subscribers. The basis of the channel had to be some form of local news because that was unique. Joyce did a deal with Radio Clyde where, for a fee, they revealed their stories in the morning, the cable station sent its cameraman to join the radio reporter, who would then also do a piece to camera. She soon found that they could source a lot of their own stories, especially ones which directly affected the area they were actually serving. A story about the police fixing locks to pensioners' houses for nothing created goodwill with the police, who then fed Clyde Cable with new stories, and with pensioners who were thus encouraged to offer more information.

Rather than a daily news magazine they produced news headlines from the studio twice daily with a full news magazine programme twice weekly. A schedule of acquired programmes was constructed around the news, with text information during the off-peak periods of the day. Films and tapes of local interest programming came from the university, transport authority, tourist board, gas board, RSPB, and whisky distillers.

The aim was to give the channel a unique local flavour, to position it as a service that could not be seen unless you had cable. Marketing got quite involved - suggesting local events which could be filmed and sales people would come back with ideas for stories. They launched local bands, one or two of whom have gone on to professional contracts, and this strand of programming has been popular throughout the life of the channel.

Clyde was also the first of the new cable stations to take the issue of community access seriously. It was a conscious decision to make access a separate channel. Much of the output came from a videobox moved around in the community. It started in a local library and at first people were not very sure about what to do even though there were instructions. There was no monitor so it confused people but children quickly took to it and learned that it had a two-minute time-out. There were lots of inventive ways of using the videobox. Children would bring their own pets, or stray kittens, used it for selling their toys or something they had found. Adults talked about local problems, there were singsongs, and a long-running genre of contribution was Elvis imitations. Of course there was obscenity as well, which was cut out in the editing of the package before transmission.

Access programming relied on a U-matic camera, recorder and edit system which could be borrowed. The Clyde Cable team anticipated that a lot of community groups would make worthy lobbying type videos, but what they got was youngsters who made drama, satire and comedy. It became a seedbed of talent. Some of the access programmers went on to make community programmes with the staff team of 4 or 5 and some have gone further into professional production.

Like all the other systems Joyce Taylor was under continuous financial pressure. The unit made videos for the police and local authority for cash, programmes which they could show as well. The problem was that the more videos they made the fewer programmes of their own were completed and it became a standing invitation to separate the unit as an independent production company.

Advertising was difficult. The network reached too few subscribers and the cost of sale was not worth the return gained. Small retailers do not offer a serious market and agencies were not interested in 4000 subscribers.

The harsh reality of funding came when the system's backers ran out of investment cash. Cuts started, the access channel ceased and the local channel became leaner with less. Joyce Taylor left for United Artists Programming in March 1989 and her sales manager, David Drysdale, took over. He departed in 1991 and after further slimming the local channel ceased making new productions in early 1993.

Croydon

The fourth of the new build systems to get underway was Croydon, also in October 1985. The founding managing director, Alan Robinson, stated at the time that he was "loath to serve up what cable has been criticised for" in the USA. Fraser Ashford was the one member of staff dedicated to programme making and was there with his camera in the home of the first subscriber when the cable connection was installed.

Fraser had started his involvement in programming for Croydon even before the franchise was awarded. Alan Robinson was backed by father and son Victor and Matt Oristano from the USA (they also founded the Bravo channel). Together they submitted one of the more imaginative applications for the pilot franchises and when they won the licence it was natural that Fraser Ashford and the programme makers who worked with him should be asked to join the cable company.

Crystalvision was founded as a production company associated with though not wholly owned by Croydon Cable. Its ambition was to provide local programming for the whole of London based on Croydon and possibly, one day, being able to use the nearby Crystal Palace TV transmitter site.

Another motive behind founding a production company to run a local channel was that it would also act as an independent head-end operator which could insulate the cable company against any militant union action. Such action never arrived and eventually the head-end operation was brought in-house and Crystalvision became totally independent in 1988.

The first programmes in 1985 were daily news bulletins made in the offices of The Croydon Advertiser. Using the newspaper's sources the news was read to camera by one of the its journalists, Roger Bing. On his way back to the cable company office the cameraman shot any 'wallpaper' shots required to illustrate the news, and dropped in on anyone who had made a controversial statement to get a full quote from them on tape.

In 1990 United Artists, who had taken over the Croydon system and won the adjoining franchises of Sutton/Merton and Kingston/Richmond, decided that a local new/magazine programme and local sports coverage would add value to their network. Crystalvision was invited to work with United Artists and Cable Today was born. Crystalvision produced up to 4 hours a week with 5/6 people for nearly three years before circumstances forced United Artists to bring the programmes in-house in the late summer of 1993.

The news provision arrangement with the Croydon Advertiser has continued, although it now comprises a formal news bulletin shot in United Artists own studio, and is only illustrated if shots can quickly and easily obtained. Cable Today is usually pre-recorded, sometimes several days in advance, but one continuing daily task is to insert the day's news bulletin. These programmes cost United Artists London South about £250,000 per annum - a price per hour which no broadcaster could possibly match. Through a combination of commitment, multi-skilled staff and careful planning UA producer Deborah Ward and her team are the most prolific programme makers in UK local cable television.

Cost Lessons

These early efforts had a concentration on news. Research into audience wants in local television show that local news is the one local service which customers clearly identify as one they want and which is unique to cable. There are many other categories of local programming which they like the idea of, particularly sport, but local news is consistently high in the wanted list.

Running a news operation is not a cheap option, even if the programmes are live and light on final editing demands. The nature of news coverage is demanding in manpower and people are costly. News magazine programmes offer a low-cost way of being topical - studio interviews (once you have a studio) are just about the cheapest form of programming you can make and never cheaper than when the interviewees are seeking publicity.

But such programmes are topical and the material has little repeat potential within hours. Two or three repeats during an evening of a local news programme can extract value from the investment, and a weekly news 'round-up' offers another route.

These early experiences demonstrate that the practical business issue is not getting hours of programming from the investment, but scaling the investment so that costs can be correlated to the revenue, or other benefits, that local news can generate. While audience numbers are low a purely financial equation is impossible. Television news, perhaps unfortunately for the local context, cannot be sponsored. Even though linking with other local news organisations has proved to be an effective route to controlling news gathering costs the experiences of Clyde and Coventry demonstrate that there are stresses in the practical relationship.

Broadcast standard equipment was chosen by Aberdeen and Coventry, industrial grade kit by Clyde and Croydon. The capital cost differences were huge. While these costs have come down (and the differential has also shrunk) the size of the investment in a news operation, and its on-going costs in terms of manpower, have dominated thinking about local television to the inevitable detriment of the opportunity.

1989 was the year when the second generation of local programming looked its weakest. The new Croydon programme had not yet started, Coventry had been the last of the early news magazine programmes to close and although Clyde was still making regular programmes with a team who also ran the head-end, there had been no new channel start-ups. Few franchises started operations during the later 80s.

The key newcomer to local programming in 1990 was East

Lancashire Cablevision. The Maclean Hunter franchise invested in a 3-camera OB truck, studio space (a warehouse used on a drive-in basis), an editing and transmission area and a further location production unit. They recruited a staff team and brought in a local radio journalist and presenter, Sally Moon (Sally Allen), to head the start-up of Channel 7.

In recognising the failure of the broadcaster models of local television in the early broadband franchises there was a belief amongst the few actively involved in local television that Maclean Hunter could represent the breakthrough that had long been sought. The Canadian model of local TV was better understood in the UK than in the early 1980s, and British television generally was changing. The elitest broadcaster approach, from which local television was struggling to free itself, was also under threat in its own back yard. Never was this more apparent than when Granada TV made a series about multi-channel TV in the Lancashire village of Waddington, just a few miles from Maclean Hunter's Blackburn base.

The Waddington Experiment

In terms of gaining publicity and of diverting attention from the real activity of local television Waddington was superb. Granada Television worked with Salford College of Technology and Lancashire Polytechnic to set up a cable television exemplar system in the Ribble Valley village in the spring of 1989. The idea was to examine how a small community would respond to multi-channel TV and then to report on it in a major TV documentary series for Channel 4. Almost as an afterthought the programme makers decided to include a locally produced village channel in the programme line-up. Inevitably WVTV, the village channel, went on to capture all the headlines.

It was decided to offer 10% of the village's households a multi-channel service for one month. The 32 households finally chosen represented a cross section of the demographic make-up of the village population. An important part of the research examined overall viewing habits and the monitoring equipment counted the number of 'zaps' on the remote controls from each household each day. It is interesting to note that viewers criticised the number of repeats and the poor quality of what they readily recognised as 'low budget' programmes on the satellite delivered channels offered to them.

So that the 'village channel' WVTV, could be received by all the village households - not just the 32 cabled homes - a special dispensation was given by the IBA for a low-powered local UHF transmitter to be set up. Thus WVTV Waddington, for its one month of life, was the first, and will possibly be the only, local TV channel ever to be terrestrially broadcast in the UK. Its 100% reach is also virtually unique (only Milton Keynes can match that figure).

WVTV went on air for about an hour Monday to Friday for four weeks. Production started as a joint operation between college media studies students and the villagers, with the villagers taking the on-screen roles. It started very conventionally, largely because the team felt they had to offer a similar look and feel to established TV channels, but it soon acquired a style of its own. The village butcher's wife fronted a weekly cookery programme, during which she cut up whole carcasses, prepared various dishes, and even managed to burn a lamb joint. She was a hit. The joints she prepared on air were sold afterwards in the village butcher's shop.

A local child's pet goldfish became the between programme intermission ident, the camera team were thrown out of a parish council meeting, and the subject of skateboarding in the main street, against police advice, became a matter of great debate.

Each Saturday morning WVTV became WCTV - the C being for Children. Everyone who wanted to be a local star got on air, even if it was on the phone-in using the single telephone line. It brought the extraordianry sight of primary school children wandering round the village with S-VHS camcorders and microphones seeking someone to talk to, or to setup some stunt to watch. Those involved happily admit that it wasn't what professional programmers would have made but the freshness, pace and camera wobbles produced a programme which was even described by professionals in the Granada team as 'compelling'.

An important factor in the success of the community TV activities was the live programming. In the last week of the experiment a full charity telethon was conducted, involving over 100 local people. The energy of the local people involved is highlighted by the approach to fund-raising to pay the costs of hiring the village hall as a studio base (£176 a week). Without any reference to the IBA (then the broadcast TV industry regulator and under whose aegis the experiment was being run) airtime was sold and adverts produced and presented by the regular presenters whose programmes the ads would be inserted into. This naive breach of the advertising regulations was only the first to tax the regulators. One presenter, a solicitor's clerk, actually made an advert for her employer - only days after solicitors nationally were first allowed to advertise. It was a national first, and added to the headaches faced by IBA officers when their attention was drawn to this enterprising activity.

The Waddington experiment could hardly be called a typical example of local television. It worked with loaned equipment, had a high level of unpaid semi-professional support, and could concentrate its effort, exhausting creative ideas knowing that after the month was up there would be no need to sustain the service.

But it did demonstrate one thing that enthusiasts for local TV have never doubted, but which cable companies have shied away from - the fact that local people can make local programmes which local people want to watch.

CHAPTER 4

FRANCHISE PROMISES

There is no hyperbole like that of the cable franchise applicant

The 1984 Cable and Broadcasting Act required cable franchise applicants to indicate their plans for local and educational services. This reflected the emphasis placed on the potential for such services during the political debates of the early 1980s.

Guidance notes from The Cable Authority indicated what they would be looking for in an application, and what confirmation of these proposals would be sought in interviews with both officers and members. The formal application interview with the full Authority in session also tackled the subject of programming and did so, sometimes in a very probing way.

I well remember the interview for Camden when Cable Camden's local programming plans were subjected to minute scrutiny on the subject of editorial control. The structure offered included remarks about establishing collaboration with local groups. At that time, Camden was a hot-bed of social activism with a very left-wing local authority. TV take over by militant activists was a political fear and despite the executive powers the Authority held which could prevent such an event, firm assurances about editorial control were necessary. Those of us who faced more than one of these interviews soon realised that any suggestion of 'access' however tight the controls which might be intended, was best omitted (even if considered in the first place).

Emphasis on education, music and sport was safer territory. A clear statement on how editorial control would be maintained and how the Cable Authority's Programme Code would be implemented (with particular regard to 'due impartiality' and 'avoiding undue prominence' for particular views and opinions) were an important part of the franchise bid presentation. The Authority's officers and members would otherwise be certain to want these policies to be clearly stated during the interview process.

Any public domain is populated by promises and it is usually easy to see the ones which are being fulfilled. The commercial reality of cable in the early 1990s has fostered an opportunistic blindness to cable franchise promises. It is easy to read the local and educational programme promises of successful applicants of the mid- and late-1980s and realise that the promises are not being delivered. But it is just as likely that the franchisee today has little remaining relationship with the management team which won the bid in the first place.

How seriously the Cable Authority took these programming promises can only be guessed at. The main theme of the award process was about trying to establish a viable industry. Economic and funding factors dominated. A non-viable business plan would not get a franchise regardless of any high-flown programming promises. Equally a company with a sound business plan and no local programming promises was not generally disadvantaged unless a better business plan with programming promises was presented.

Jon Davey is tactfully silent when asked about the effect that promises actually had in the final award of particular franchises. The Authority's former Controller of Programmes, Tony Currie, maintains that his scepticism of some of the promises of the 1980s is only matched by his sense of disappointment today that much of what he knew was practicable, and thought was being honourably offered, has not been delivered.

Whatever the truth of the matter the substance of agreements reached between successful cable franchise applicants and the Cable Authority during the awards process is now a licence condition. The PDSL licence from the ITC which most of the successful franchise applicants now hold allows these promises to be written in as a Statement of Local Services. However, the sad state of the law is that the ITC only has one sanction if these conditions are not complied with - and that is revocation of the licence. While investors are still consolidating their new businesses, developing their medium/long term plans and building their networks, there is little likelihood that the ITC will even consider embarking on an enforcement process for a local programming 'obligation'.

What did they Offer ?

It is an intriguing question to consider to what extent the many US acquirers of UK cable franchise businesses have examined the promises to which they may one day be held as a condition of operating these businesses, and which, in theory at any rate, threaten their operating licences. Whether their shareholders in the USA would be happy with a failure to disclose such threats, or a management failure to fulfil potentially licence threatening obligations, is a matter of conjecture.

Consider Southwestern Bell and its partner Cox Communications, which have inherited some of the most outrageous public promises to local television in franchises they now operate. West Midlands Cable Communications, applying for the Black Country franchise declared:

> "*Five local interest channels will be provided ...*"

The application devotes more than three pages to Black Country Television Ltd, just one of its proposed local channels. BCTV "*will be a commercially-supported television station*". Amongst the strands of programming on offer will be:

> Local Sports - "*building on direct experience of elsewhere, the Company will also develop ties with leading sporting bodies ...*"
>
> Local Schools - "*a member of the production staff of WMCC will spend the first 6 months of the second year of the franchise term meeting with ... to design a yearly schedule of sports and drama coverage.*"
>
> The Neighbourhood Line - "*A special series designed to cover local festivals and activities within each of the region's parliamentary constituencies ... an umbrella for special events.*"
>
> Multi-Cultural Programming - "*...promote WMCC's priority training classes for minority residents interested in TV production.*"
>
> Woman Today - "*... career and life options available to women nationally and within the franchise area.*"

Age Link - for the over 65s - "... *a broad spectrum of practical issues ... will also explore the leisure and entertainment opportunities of interest to over 65s.*"

Speak Out - a live 30 minute discussion programme with phone-in which "*will be controlled in large part by both the speakers and the subscribers at home.*"

In addition a Public/Educational Access Channel was to be launched with facilities provided by the company. The princely sums of:

£10,000 in year 1
£25,000 in year 2
£15,000 in year 3

were allocated to acquisition equipment and a further £15,000 to £20,000 a year "*had been allocated to tape budgets*". WMCC also stated that it would provide technical and administrative staff, and offer free training. There was also to be:

Exchange/Home Shopping using photo-videotext plus a daily home shopping service
Black Country Focus - a community noticeboard
Black Country Jobs - a listing service
Black Country News and Business - using Reuters service
Black Country Weather - from the Weather Department.

Many of these services would no doubt have shared a channel rather than being exclusively available 24 hours a day. Without examination of the business plan elements of the bid, which were and which remain confidential to the ITC, it is impossible to know how all these services were to be funded.

In his favour one must say that the writer did expect the company to invest in some programming - however modestly. - though with the few figures shown above one questions his awareness of the costs of developing a sustainable TV production operation even when industrial grade equipment and volunteers are used. If his ability to sustain a single channel, let alone five, on the budgets indicated publicly was ever questioned remains a secret locked in the files.

Contrast this Black Country enthusiasm with the heavily proviso'd, but nonetheless specific, statements of the successful applicant for Liverpool North. Where WMCC enthused about the qualifications of its directors in managing TV production and their wish to put petty cash in to support their experience, Cable Communications waxed lyrical about the existing skill base in Merseyside and its suitability to manage local video training and production from distributed centres about the city.

"Liverpool Text and Seft Text", it declares, *"will be available from the start of transmission."* And to ensure "*wide appreciation*" monitors would be installed in public areas, railway stations, shopping centres etc.

"Focus Liverpool and Focus Sefton, will be launched when there are some 10,000 customers connected, or when there is sufficient local material regularly available to start the channels." (Neat side-step that, they avoided adding "if ever").

A later stage of development will entail a full-time in-house employee. Five educational and community centres are listed as being capable of contributing - although there is no mention of their ability to deliver anything other than occasional programmes. As if to rub it in Cable Communications publicly agreed in the document with quoted statements from one unnamed local arts organiser that:

"local cable channels could help redress the problems associated with a diminished cultural base by providing opportunities to develop cultural, social and political awareness in Sefton."

High sounding words - they may have been in tune with the political thrust behind the development of cable but were they ever realistic ? The final stage of a massively expansionist programme in Liverpool was to be:

"... the establishment of TV Merseyside (TVM). This will be a fully fledged professionally produced local channel developed in conjunction with established TV and video production companies in Merseyside"

and in addition

"Cable Communications also intends to establish a range of educational and training programmes in conjunction with local educational establishments to form part of the programming on TVM.... It is expected to launch this third stage of local programming when the cable installation phase is complete or earlier if subscriber numbers

warrant."

And the document goes on to detail that this would be in year 3 rather than year 5 if all of the company's franchise applications in the geographical area were granted. A conditional statement which gave them a get out.

Herein lay the catch. Offer me what I want and I'll play along with your requirements and, even then, I'll place such conditions on it that I may never have to do it at any rate ! This is the most cynical kind of franchise application and while it may have some weight with local authorities and others consulted during the franchising process, The Cable Authority became used to attampts to sway their judgement using such offers, and discounted them.

Attempts at Undue Influence

Any suggestion that franchise applications could be conditionally linked in this way was openly rejected by the Authority. The applications for each franchise were considered without such external factors being taken into account, although in the last months of The Cable Authority when it was clear that franchise grouping was an important way of creating blocks with coherent business plans which could attract the funding that the industry needed, applications extending an existing franchise did have some advantages over applications which would have created further small isolated franchises.

Most applicants were either more realistic in expressing their ambitions, or could call on appropriate American experience to give themselves programming credibility. In the last months of the franchising activity they are as numerous as those whose applications are more questionable. Two good examples of this more practical approach are:

"We do not plan to impose a DIAMOND conceived channel on the market.... Diamond will sponsor a community television workshop with the aim of developing programming of local interest."

"Two features Maclean Hunter cable tv in Canada is particularly proud of are its 20 year record of originating local progs even in systems with as few as 2000 subscribers and its in-house expertise which is used in Maclean Hunter's own systems and also by other

cable operators throughout North America. The Maclean Hunter principle is to have local training, local purchasing and local recruitment policies wherever possible ..."

Broken Promises ?

The issue of franchise promises on local television is possibly the biggest single sense of grievance felt about cable in the active community based media sector. Many people contributed to plans with ideas and a sense of commitment, usually at no cost to the franchise applicant, only to see their support rejected once the application they had supported was awarded the franchise.

The community and workshop based media sector went through a period of growth and prosperity in the early 1980s, based largely on joint efforts by Channel 4 and the British Film Institute to create a third force independent sector. In the major cities these groups often had a contribution to make to ideas about local programming, even if it was only informal discussion. They had every reason to hope that cable would offer them new growth. This sector is now disappearing. Grant funding has all but dried up and Channel 4's priorities lie elsewhere. Those groups which could not find a commercial footing have largely disappeared.

Those who feel aggrieved, and there are a good number, often have routes to important sources of influence. But the brutal fact of life is that there is little or no action that can be taken if the cable operator does not want to run a local channel. Even the prospect of 'third party' local channels (ie. with no active involvement from the cable operator other than carriage) is proving slow to be realised - partly because the economically viable size of audience is far higher than was ever considered during the 1980s and partly because most of the approaches made from the workshop sector are sufficiently insubstantial to be easily dismissed.

Local authorities, if they feel they have been 'betrayed' (and that term has been used to me by more than one senior local government officer) can put all kinds of obstacles in the way of the development of the cable network. At least three London local authorities have taken this approach (though not exclusively for programming reasons) and there are others elsewhere. The irony is, of course, that where this happens the population of that area are unable to take advantage of alternative telephony services, and are forced into use of satellite dishes if they want to receive multi-channel TV until such time as the cable company negotiates its way in. As a perverse subset of this approach the local authority has no powers over domestic satellite systems which meet the basic installation regulations and thus they are by-passed from the decision making processes. The cable company may suffer for a while, but it is reasonable to expect that the cable 'movement' will gather sufficient pace over time for otherwise intransigent authorities to modify their stance. Of course the promise of local programming could be used to help soften the blow. If it is used in this way I would suggest that local authorities, having once been let down, will want to make it a contractual obligation in their relationship with the cable company.

The next generation of franchises to be awarded under the Local Delivery Operation (LDO) licence structure in the 1990 Broadcasting Act will have no programming obligations, and no plans need to be stated in the applications. In addition local authorities will have minimal consultative rights. In a classical British 'fair play' situation those cable franchises which were awarded before the Broadcasting Bill (which became the 1990 Act) was published were allowed to convert their licences to LDO status shortly after the 1990 Act came into force. Although the main reasons for conversion were to do with technology change and with licence renewal conditions, it is an interesting reflection that the nine franchises which did change their licence to an LDO now have no programming obligations. These include Aberdeen (which closed its programming unit shortly after starting up and is now fully built), East Lancashire (whose programming has been closed following the takeover from Maclean Hunter by NYNEX) and Croydon (also fully built but with one of the more extensive local programming operations).

None of these franchises has any further need to pay any attention to its original local programming promises. In these franchises, as in all future franchises, the push for local TV must come for better reasons than a sense of obligation. Given the impracticability of enforcement in those areas which do have obligations (even when the details of the actual obligation are somewhat unclear) I would suggest that enthusiasts for local TV have to promote the same sound, commercial reasons for a local channel as will be applied by every future LDO licencee.

The case for local television has to be made rationally.

FRANCHISE APPLICATION CASE STUDY

CABLE LONDON

I like to think that the applications for the four north London franchises which now make up Cable London's area of operations were amongst the best thought through in presenting local TV in franchise applications.

Even allowing for some sense of 'selling' the idea the Hackney & Islington and the Enfield franchise applications, which were considered by the Cable Authority during early 1990, demonstrated an incremental plan covering the whole franchise block for which licences were being sought.

The plan was based on creating something simple to run and cheap to operate, while background research suggested that full commercial operation at some time in the development would be a viable option.

The Hackney & Islington application started by describing text channels and how a wide variety of local information needs could be met. It then outlined a policy for growth of programming offering *"a clear development path into a full North London channel"*. It refers to evaluating progress, creating a justification case for the channel's continuing existence, and goes on to position topical information as the key programming opportunity. The build up was graphically illustrated (see overleaf) and this part of the application concluded with the following paragraphs:

The first priority is to establish the local identity and develop the ability to provide locally relevant programme services.

This can be quickly achieved through a "text channel". This will operate 24 hours a day and will incorporate local information (leisure, arts, community groups, sports, welfare and council news), news and travel advice, and special sections on such matters as eating out, children's health and so forth. This service will be developed and initially run by Cable London.

The proposed development path calls for training of community volunteers, followed by production of local documentary and discussion programmes which can be transmitted within the text channel at specific times.

A more regular programme transmission pattern can be established once a "critical mass" of programme making skills and ideas has been established, leading onto daily programming.

Figure 3:1 Community Services

Condition	Service	Requirements
10,000+ homes passed	Rolling text	Low investment Manpower requirements small
50%+ build with growing sales and good research from programmes already shown	Some programmes	Some camera equipment Availability of editing Established training scheme
80-100% build and clear proof of retention of trained skills within area	A pattern of programmes	Infrastructure in place Trained pool of effort
	Regular programmes	Studio and associated overheads established
Full build and increasing penetration with evidence of static or reducing churn	Daily schedule	Expansion of facilities
As and when opportunities arise	Special events	OB capability
	Programme sales/ exchange	Sales/marketing structure
At a future time should the right mix of opportunity and requisite skills arise	Technically complex programming	Innovation skills and investment

And now the Enfield application:

3.4 Local and Community Programme Services

Cable London has always made a commitment to the provision of locally generated programming for the community in its franchised areas. This provision is based on:

a clear strategy
the commitment of resources
on-going local consultation

We envisage that in a four to five year time-scale as many as four local television channels will be available to the community, and in addition ethnic/language groups may well have their own dedicated channels. We see three key elements in the development of this scale of service.

Most of these channels will be text based, some will be truly local (ie within a specific borough) and some will be of wider interest (ie the whole North London community). It should be noted that today's technology means that a text channel can incorporate high quality graphics and photographic still pictures. Such channels form the core of our local and community strategy in the early days.

There is a great wish within the community to be involved in programme making. This desire generally exceeds the abilities of the community to handle the technical, creative and organisation issues of television. Thus the second key element of our approach is to encourage the development of skills, and through training, to enable community groups to undertake programme making.

We have identified, and market research confirms, that a North London TV channel carrying local television programming will be attractive to the North London community encompassed by Camden, Hackney and Islington in the south, Haringey and Enfield in the north. To develop such a channel, working alongside appropriate programme makers, is the third element, and a long term commitment, in our overall programming strategy.

Programming Development Plan

The development of local and community programming is dependent on a number of factors. Briefly these are:

number of homes passed by cable
growth of homes connected to cable television
commitment of the community to cable programming

The first two determine the nature of the operation, the latter is the major determining factor for the scale of the operation.

The "North London Channel" will contain a programming mix which reflects the North London community and encourages debate, the development of artistic and musical talent, promotes local activities and contributes to community life. Much programming will be live, and whilst viewers like regular schedules, there will be a preparedness to adapt to the event, rather than always be driven by the clock.

If Cable London is successful in both its current applications (Hackney & Islington and Enfield) we expect to reach 20-25 hours per week of original material within three years.

The rich diversity of cultural life in North London will be reflected by the channel. 'Club' music is a commonly expressed need by younger aged groups; black, Greek and Asian theatre groups hold regular programmes; and various festivals provide a focus which enable groups to communicate and to inform and educate on a wider platform.

Taken as a whole, the five Boroughs that make up our North London strategy include sizable communities that are large enough to attract common interests and involvement. Based on the 1981 Census figures, which underestimate the size of these communities, the total area includes a population of nearly

88,000 people from European and Mediterranean countries, primarily Cyprus, 56,000 from Afro-Caribbean backgrounds, and 28,600 from Asian communities. The opportunity will exist within the "North London Channel" for community based programming of interest to the whole area. This may be ethnic in nature, and we can anticipate that a community might undertake the programming of a whole evening in order to reach out to its members spread across the North London area.

Cable London has begun discussion with independent producers about how this service could be structured and operated to the benefit of the community.

Summary

Our commitment to local programming will offer the community better access to news and information and the opportunity to express its voice in new ways. The full range of our planned services will provide a new focus for community groups, will allow individuals to develop new skills, and will broaden the base of entertainment available through television in Enfield and the rest of North London.

CHAPTER 5

INTO THE 1990s

One step forward two steps back

From all that has gone before the uninitiated reader might think that local television was a bubble which by 1990 had burst. At the time of writing (early 1994) this 'third force' in British television is far from dead and its roller coaster of activity could continue for much of the year. There is sufficient positive activity to give good reason to believe that long lasting local television initiatives based on real commitment by cable operators will soon be underway.

When franchising finished under The Cable Authority during 1990 and the UK TV industry began its run to the millennium under the newly formed ITC (which subsumed the Cable Authority from January 1st 1991) the signs for local television were positive, even if there was no evidence of widespread cable company enthusiasm. There were doubts about whether promises made at the time of franchise applications would be honoured, but the same could be said about funding commitments, and seeing these commitments met was far more critical to the future of the industry.

Swindon, Croydon, Clyde and Coventry came into the 90s with channels on air and active production work underway. Windsor and Westminster were showing local information and carrying local advertising (including TV ads) and a wide variety of cable operators carried simple local noticeboards running from text and graphic computers. East Lancashire Cablevision had come on stream during 1990 bringing the full force of Canadian style local TV into the UK for the first time. Cable London opened up its local channel in 1990 and Hellenic TV was putting its plans into place at the end of the year. But progress since then has been stop-go and two of these early leaders disappeared during 1993. To account for the continuing ups and downs it is important to gain a perspective of how the UK cable industry as a whole changed during the years from early 1990 to late 1993.

In early 1990 the cable industry entered the third clear phase of its development. This was when North American corporations became heavily committed to acquiring substantial interests in UK franchises and started to build the networks they had already acquired either through joint ventures or nominee companies. Until the 1990 Broadcasting Act came into force non-European companies were not allowed to control UK cable companies. This had been evaded by setting up offshore trusts and foregoing formal voting rights. The 1990 Act corrected this anomaly but until such time as they could install their own managements the US companies played their investments in a cautious way.

There were also some very good practical reasons for caution. It was already evident that technology change during the 1990s would allow new systems to be built with vastly enhanced capabilities, and the UK's invitation of a largely de-regulated telephone market offered promising potential. Until the Government's Duopoly Review was completed in March 1991 most investors controlled their enthusiasm. They had bought their way into the market, and in doing so had acquired obligations. But they sought a free hand in developing the combined networks that technology promised and that meant constraint on BT and its monopoly powers. When they finally got the market conditions they wanted there were still some doubts over market performance, and in particular the acceptance level that alternative telephone services would have in the residential market.

In 1990 US West (one of the controlling shareholders) advised Cable London to assume a 10% penetration level for telephony and to build for 20%. Within weeks of first marketing the new service capacity had been sold out and demand was still growing. It was an expensive mistake for Cable London and for Mercury, but better made earlier than later. For Cable London it meant a rebuild and re-plan of network design, but the industry had its first real indicator of consumer enthusiasm which has been born out in every area since where cable telephony has been introduced.

By the end of 1991 the recession began to bite in both the USA and UK. The national cable TV penetration figures were reluctant to rise much above 20% although some franchises were doing notably better and the indicators for telephony were highly promising. But there was still plenty of scepticism. The sole objective of the operating franchises was to build as much network as possible, to as high a quality as possible, while consuming as little cash as possible. This was not an environment which encouraged an investment in local TV and could allow it to gestate, grow and flourish. With insignificant audiences there was no revenue potential from advertising, and cost control was a financial imperative.

January 1992 demonstrated the first spurt of growth of cable telephony. It gave the evidence that was needed to encourage a year of concentrated activity in a small number of places. The results were so promising that by the autumn it was clear that the nature of the industry was changing dramatically. A year later, by autumn 1993, more than half the operating franchises had launched telephone services and many of them had committed the capital investment required to undertake their own switching. It was a complete about turn. As telephony penetration rose from insignificance to a level which approached that of cable TV the old entertainment led emphasis of the cable industry had gone for ever.

At the end of 1993 cable TV penetration was still languishing at around 20%. During the year the number of homes passed by broadband cable increased by more than 40%, a total equivalent to that achieved in the whole of the first six years of broadband cable build activity. By the year end most cable franchises had the promise of full funding and a fourth stage in the growth of cable began.

Local TV in the Early 90s

There was an air of great optimism when 1991 dawned and the ITC took over the regulation of cable TV. With so many promises for local channels now 'in place' there was good reason to believe that a new programming era was about to dawn.

It is only an estimate but accounting for the multiple local channels offered in many successful franchise applications there was the expectation of as many as 200 locally originated TV channels of one kind or another by the end of the 1990s.

Probably no more than half of these promised actual moving image television and there was little public indication of how this level of activity was going to be funded, other than vague statements from successful franchise applicants about 'commercial' objectives. Those aware of the optimism of 1982 reports on the prospects for self-funding local TV might have had some sense of deja vu.

The leading newcomers amongst the local channel operators were East Lancashire Cablevision (in Blackburn) and Cable London. Between them they typified the pattern for the next three years. Cable London began with a simple 'photo-videotext' system although the enthusiasm shown by local people for developing a volunteer group led to an acceleration of its plans outlined earlier in Chapter 4.

East Lancashire Cablevision was owned by Maclean Hunter and the proof of the Canadian company's franchise promises soon came in the form of an extensive investment in production and post-production hardware, and staffing. Cable London was typical of the low-cost start-up with ambitions, East Lancs, because of its investment in staff and equipment, set the example which others could only envy.

The Guild of Local TV

In December 1990 the Institute of Local Television held its "Second Really New Television" Conference in Edinburgh. This event was to mark a transition in the affairs of local television. Coinciding as it did with the imminent demise of the Cable Authority and the appearance of the new television regulatory regime under the ITC it could hardly have been more timely.

The conference spent its first day raising critical comment about inactivity in the cable industry, and despite a presentation by Hugo Davenport of Cambridge Cable on the economics of running a cable company allowed no hearing for the funding problems that the cable industry had been facing. The lunch period on the second day was billed as the inaugural meeting of the Association of Local Television Operators (ALTO) membership of which was proposed for cable companies at £1000 per annum. It was also proposed that ALTO would be run by the Institute but, with its sponsored conference taking a position strongly critical of cable, the cable representatives present boycotted the idea and decided to take steps to set up their own association.

Thus dawned the Guild of Local Television - born a year later after three conferences in Glasgow, Leeds and Birmingham. It was formally instituted in Cable London's conference room in Camden Town in January 1992 with 27 members present. These members represented independent local channels, producers, education colleges, one broadcaster (YTV) was present and of course cable operators. This catholic mix remains typical of the membership today and of all the bodies set up to represent local television the Guild can probably claim to have the widest membership base and the broadest objectives.

But despite the optimism that forming a new organisation creates the fact was that the cable industry was in deep gloom and the prospects for local TV were, at best, hesitant. Funding for many of the smaller companies committed to UK cable had dried up as the revenues of parent corporations dived in recession hit North America. Even those companies which kept building were not moving forward as fast as they really wished, and every spending proposal was greeted by shareholders with extreme caution. Local television, high in the priorities when applying for a franchise, was identified as a straight cost - offering no real rewards, and in the absence of a regulatory regime which could demand that promises were met progress was painfully slow.

Local Television Progresses

Even those operations which were underway moved on with great caution. Coventry, long one of the flagships, had changed hands with Devanha taking over operation of services on the network from BT in late 1991. Although the television production equipment owned by the company was six years old in autumn 1991 it was clear that if the channel was to continue transmitting it would have to continue working. Such were the realities.

In January 1992 Sally Allen, the founding manager of the East Lancs Channel 7, was poached by Windsor TV. Her new channel opened up in the late summer of 1992. At about the same time Leicester Communications started its local channel, also adopting a Canadian model of organisation heavily reliant on volunteer effort. Cable London's local channel, CLTV Grapevine, came out from behind its 100% photo-videotext beginnings to start transmitting programmes in early 1992, again from a largely volunteer production base. And down in West Glamorgan Starvision Network began a daily system and network news programme for its subscribers through the enthusiasm of marketing manager Nigel Griffiths.

In the same timescale the first independent (or 'third party') local channels got underway - Hellenic TV in north London and Open Campus TV (OCTV), the Blackburn College based educational channel. In early 1993 Videotron began a London multi-cultural channel, transmitting up to five hours of independently produced programming each weekend, and Wire TV, the cable operator funded exclusive national channel (managed by United Artists) began transmission in May 1993.

By the end of the year there was programming in Hemel Hempstead - through an independent volunteer group, Parallel Pictures; and Birmingham, now the franchise with the largest number of subscribers, was transmitting Channel 021. It shows a daily schedule of local interest programming, some of it specifically produced, wrapped round with local recorded music, and news and topical interest material in photo-videotext form.

Yet sadly the Clyde Cablevision channel ceased producing new programmes in early 1993, and by the end of the year Channel 7 in East Lancashire and Grapevine in Cable London had ceased operating for reasons determined by their cable operator managers.

Looking Forward

But if the number of TV operations was slow to grow the number of local channels using image and text systems to provide local information, and the number of franchises beginning to take advertising seriously was increasing.

The belief is that channels with text based systems will find the opportunity to attract advertising as their audience numbers grow. This, they hope will pull an ad sales operation into existence, and inevitably local programming will be needed to help sustain a meaningful local commercial channel. Is this pure optimism ?

The platform for the later 1990s seems to be set. There is no reason to believe that local channels will not continue to come and go, but the trend will be a steady increase in numbers. Insertion of local advertising on the channels which allow them will become a major activity for cable companies. United Artists (TeleWest), Birmingham Cable and Northwest Cable in Preston are the first to establish such operations and their results are promising. These advertising moves may have an impact on the introduction of local channels.

Resolving the business issues and the organisational matters which arise in taking advantage of this mix of opportunities is what the latter part of this book is about. There are no clear formulae. It is already apparent that there are many lessons to be learned from the past of local TV in the UK, and there are many more to be explored - some of which I believe we can already identify through projecting past experience.

The third force in British television may have had a hesitant birth and a staccato adolescence, but the road for its progress into maturity is now mapped out. Our hope is that we can help cut out some of the twists and curves on that road and offer an easier path into local TV for cable companies, producers and independent channel operators alike.

CHAPTER 6

WHAT ABOUT THE AUDIENCE ?

Do they really want it ?

In the world of multi-channel TV can subscribers really be bothered with local television? This question occurs again and again, sometimes as an excuse by cable operators for not getting a local channel underway, sometimes by those with local interests and wondering whether they could be funding a 'third party' channel. It has also been used in at least one case to support the argument for closing a channel.

The evidence that local television is a potential audience grabber is enormous, although empirical data based on independent research of actual examples is hard to find - largely because funding such research is not usually a budget heading which a local channel manager can get agreed by his senior management. Why cable companies are reluctant to fund specific audience research to support a channel's existence remains an unanswered question.

The Great Myth of Local Television

The greatest myth to dispel is the one which broadcasters promote - local TV is of poor quality therefore no-one will watch.

In the 1980s, as cable loomed as a threat to the traditional broadcasters, promoting this myth was a popular theme for late night programmes. They took a look at examples of local TV from the USA or Europe and openly lamented the poor quality of content, low editorial standards, questionable production values and lack of technical competence. Those adopting this highly selective approach were, of course, able to chose what clips they liked from whichever programmes they did not like in order to prove their point. As an approach it manifested some of the self-same lack of dispassionate editorial values which they were criticising.

There is little doubt that this campaign created an impression on significant elements of the British audience. To many people any mention of cable television is coupled immediately in their minds with pornography, openly biased campaigning programmes by wacky or extremist groups, on-the-street strippers, and performing dogs, cats and goldfish.

Some local television is quite clearly not what a broadcaster would expect from a fully trained and experienced staff production and technical team. The real issue, and the one the audience can identify with, is the content. Relevance, uniqueness and personal involvement will conquer any concerns an audience may feel about production values or technical standards, even if they cannot articulate these concerns. Programme quality, it may be said, is in the eye of the beholder.

Any serious concerns about editorial standards can be addressed through the regulatory structure. The ITC's Programme Code is quite specific about what can and cannot be allowed in most instances where treatment of a subject may be contentious. Even if 'taste and decency' are open to subjective interpretation there is a well established set of precedents created over the years to guide a programme maker. The fears, which largely stem from deliberate misrepresentation of local television as 'access' television US style, are absolutely groundless.

The Waddington experiment provides the best observed and most comprehensively evaluated data on the power and impact of local television. The programming (and its viewers) were more intensively watched and researched over the month long experiment than any other example of local television before or since. It is highly unlikely that in such a small community, the commitment to create the variety and excitement of the content could have been sustained for much longer than a month, but with a population of barely 1000 that is hardly surprising. That they achieved such variety at all is the key issue. There were all sorts of technical and production accidents, many of which happened live - and no-one minded. Indeed it could be argued that they increased the popularity of the channel - a factor noted in the final report on the experiment.

The 32 homes in Waddington which watched their TV on cable during the experiment provided a measurable sample of the 300 local households who could receive the channel (it was also transmitted on low-power UHF with special permission from the IBA). By the fourth week 31 of the 32 households were watching WVTV during its hour of transmission. This was at peak viewing time (7pm to 8pm) where the local channel competed with top soaps, comedy, music and topical programming on the four broadcast channels, plus the niche material on the 20 or so satellite delivered channels also available.

Waddington remains the only measured example of local television's audience when compared with other channels although, as we shall see later in this chapter, there is one other instance of independent research.

Pre- Franchise Market Research

We have already seen that many franchise applications carried out research into audience wants as part of the justification for the franchise application.

The Cable London research in Enfield (carried out by Continental Research in 1989) has been widely quoted in public and is worth repeating here:

> *The survey was carried out in January 1990. It was conducted by telephone over a weekend and weekday period, and contacted 504 respondents who were Heads of Household or spouses (these being the key decision makers on cable subscription), sampled to match demographic characteristics of the area.*
>
> *The results supported the prospect of cable television and particularly the approach and emphasis that is being adopted by Cable London.*
>
> *Respondents were given a general description of cable television services, and descriptions of sixteen channels they might expect to find on the Cable London service. They were then asked for their interest in the different channels, and their interest in subscribing to receive such services ...*
>
> *... the question asked for the level of interest in each channel, and bringing together those who responded as very or quite interested in receiving a specific channel, and who were also potential subscribers to the cable services revealed the priorities shown in ...* (overleaf)

	% Potential Subscribers @ £pm	
	£9.95	*£19.95*
The North London Channel	*88*	*89*
Discovery Channel	*87*	*81*
Sky Movies	*81*	*88*
Sky One	*77*	*90*
BSB Galaxy	*75*	*84*
Sky News	*70*	*78*
BSB Now Channel	*70*	*64*
Bravo	*68*	*71*
Superchannel	*65*	*70*
Screensport	*57*	*62*
BSB PowerStation	*57*	*61*
MTV	*55*	*59*
BSB Sports Channel	*53*	*58*
Eurosport	*51*	*56*
Lifestyle	*48*	*48*
Children's Channel	*43*	*48*

As in Hackney and Islington, the North London Channel has the greatest appeal of any channel described. While market surveys frequently find a strong interest in a local channel, this level of appeal is unusual outside the North London area.

The rating of other channels is in line with normal expectations, with the emphasis on films and broad based entertainment. Compared with other areas there is a stronger interest in the more informative channels such as Discovery and News.

In 1989 ITC Research (no connection with the Independent Television Commission) carried out a survey for a south of England franchise applicant. Asked about a community TV channel respondents were requested to place it into one of five categories - *very interested, interested, not sure, not interested, no response.* This channel received as many *very interested* votes as sport, more *interested* votes than any other category of programming, more *not sures* than any other category, slightly fewer *not interested* than new films, and the lowest number of *no response* of any category. Over all it had as high a level of *interested* votes as news and feature films, and more than sport and children's programmes.

In the younger age groups the community channel had the highest level of overall interest although feature films (and in particular new films) generally led the *very interested* category. Even for the over 65s the community channel scored well.

In 1991 Councillor Nick Brown, then leader of Derby City Council, privately commissioned a study from a media student, Maarten van der Made, as a work experience project. It was one of the most thoroughly documented independent examinations of a local television opportunity in a single locality that has been undertaken, although it did not reach out to the 'man in the street' and poll his opinions. One important comparison the writer made involved other local media, radio and newspapers.

In view of the sales figures of the Evening Telegraph there is a demand in Derby for local news. On an average, over 73,000 copies are sold every day. Trent Radio claims that 257,000 people listen to their station at least once a week. According to Radio Derby about 225,000 people listen to their radio station at least once a week.

Although the writer notes that these three also cover an audience outside the city itself they do point to a significant demand for locally provided news, information and entertainment, and they offer an easily accessible way of expressing a viewpoint back to the 'centre'.

The Competitive Arguments

At about the same time a study was undertaken by social sciences students at Brunel University looking at the potential customers of cable who do not take cable when it is offered to them. The survey was conducted in a London satellite area which was already largely cabled. The important indicators from the data are not that they confirm a demand for local television, or indeed for cable, but what can be drawn from the study because it does NOT ask the questions about local television.

A sophisticated questionnaire was developed for the interviews which were undertaken by students. They found that the strongest interest in cable was in households with no children and that, together with the age of head of household, this was the key demographic determinant in taking out a cable subscription. When it came to barriers to subscribing it was clear that more than one-third of all people approached by cable salesmen were happy with four TV channels and, whatever the attractions, were going to be very difficult to convert to cable. The second most important set of reasons concerned price, particularly amongst lower income groups. About one in five of those interviewed indicated that the unknown nature of the cable offering was an important reason for not subscribing.

There was an intensive look at the use of leisure time inside and outside the home. On average the adults surveyed spent 48 minutes a day reading a newspaper and a similar amount listening to radio. Among the under 50s playing sport was the most popular outside activity with 43%, comparing well with 16% going to the pub. Gardening, restaurants, cinema and theatre trips, visiting friends and walking also scored higher than visiting the pub.

Whilst there are few unexpected conclusions in this (unpublished) survey it serves as a very useful focus, confirming much of the well understood behaviour of the suburban household.

It also confirms a behaviour and attitude with regard to cable television which cannot be interpreted in terms of a typical American household - though there are a few similarities. The key issues revealed are not about playing sport, going to restaurants and theatres, or visiting friends. They are about attitudes to TV and the context for local information and news.

The local press in the UK lost circulation through fragmentation in the 1980s. One result is that increasingly aggressive local papers often quite adequately cover national news and in some areas are succeeding in supplanting a national daily newspaper in a proportion of households. This same mix of national and local can be heard on local radio. A hunger for local news was confirmed by a 1993 Henley Centre study which found that local news is a greater priority than regional news, itself a priority over national news, which has in turn a priority over international news.

The integration of ideas of local into what is otherwise a regionally dominated medium is also now apparent in Channel 3 (ITV). Most of the geographically larger companies now produce versions of their regional news with a stronger local flavour and content targetted at a narrower and more local audience.

The adult in the non-cabled household occupies some 90 minutes a day consuming a mix of non-TV national and local, news and entertainment, as well as an average of approximately 3.7 hours a day of TV. The problem for the TV channel developer (local or national) is how much he can shift this balance through increasing the TV share, or taking share away from some other TV programmer.

We have already seen how difficult it is for local television to be 'competitive' in traditional TV terms because of the costs of building and sustaining a distinctive quality service. Therefore the targeting has to be different, and the household share of time occupied by radio and newspapers is the obvious competitive area. As local is the sector in which there has never been effective competition from television and in every other sector it is succeeding the opportunity seems obvious.

Fostering Subscriber Loyalty

If news is desirable, but initially unattainable because of cost, there is a proportion of the local population who will respond well to the opportunities a local channel might offer. In the following chapters we take a hard look at volunteer programme making and the kinds of structure which can be employed to offer the audience something distinctive, and offer volunteers some satisfaction for their efforts.

But there has been one other area of programming which has attracted viewer involvement, in perhaps an unexpected way. In 1990 Coventry Cable began an experiment with a library of educational programmes gathered from a variety of sources, including productions from local colleges. The project was funded by a wide range of interested bodies including the National Council for Educational Technology, The Cable Authority and the education department of British Telecom. "The Learning Hour" was scheduled daily from this library and the Educational Broadcasting Services Trust (a largely BBC-funded charity) evaluated its use.

One feature of the experiment was 'repeat transmission on request'. It worked well and the EBST's study commented that educational television within the local context was a major opportunity for cable. Like so much work that was done in the early days of cable this experiment too disappeared into the files. "The Learning Hour" could be seen as an early example of interactivity in that it consciously sought involvement by the audience in the schedule, not just in outcomes from the programming.

Local television in practice has generall y been unable to contribute anything to our understanding of how the audience responds to it. The experiences we have seen show that a 'catholic' mix of programme sources and styles can offer a viable channel schedule in programming terms, always remembering the unresolved funding and financial problems.

Acquired programming, whether made specifically for cable transmission or acquired in the international television market, can take on a local flavour if presented within a distinctive local scheduling approach.

Coverage of local events (sports, fetes, school shows), discussions on local issues, regular slots from local religious leaders, coverage of pub and club music, even coverage of local council debates - all are feasible ideas, which have been undertaken to greater or lesser extent in one or more geographical areas. But, with one exception, this catalogue of work remains unevaluated in any formal sense, despite the fact that every channel manager who has ever covered them reports a positive audience response. The exception to this generalisation is outstanding and comes from United Artists London South, based in Croydon and serving five London Boroughs.

Subscriber loyalty and improved retention is the driver behind United Artists London South's investment in its daily news magazine programme 'Cable Today' and its associated local sports output. Research undertaken in March 1993 by Continental Research found the indicators which demonstrate its value.

It showed that 82% of subscribers were aware of the programmes, with 62% acknowledging that they had watched 'Cable Today'. Of those who tune in, 54% had watched the programme at least once in the previous week and 87% at least once in the last month. The responses also confirmed that local news and entertainment were particularly valued by viewers.

Sadly this is the only independent professional study of a local television service in a sizable franchise. Its results give us our first substantiated indications of the value of local television.

It all suggests that the ideal local cable channel in the UK would blend acquired information and entertainment programming, professional staff production, focussed local news services, local event coverage, and could also employ extensive volunteer effort, all to create a unique local service. The benefits the cable operator would gain are getting easier to identify. The financial commitment could be considerable and while a cable operator stands to gain the most if he funds it himself, the initial costs are enough to put many off the prospect. The key benefit for him is likely to be subscriber loyalty.

But what if a wholly commercial approach was taken, with the cable operator accepting that his benefit was the difficult to quantify, but persuasive, argument that local programming aids subscriber retention ? The commercial operation would need to be undertaken by a standalone business insulated from the cable company's financial and operational stresses. It would be a 'third party' company, locally based and building its own local revenue, local identity and audience loyalty.

I believe that this route, with the cable company possibly collaborating as an equity partner as well as a contracted bandwidth provider, can provide a way of creating a sustainable quality service which will be fleet-footed in its marketplace and from which the cable company will draw great strengths to benefit its core business.

CASE STUDY

At the 1987 Cable Television Association Convention the then Head of Programming at Clyde Cablevision, Joyce Taylor, addressed the dilemma of meeting the audience from a subscriber retention, sponsorship and advertising viewpoint.

The partnership between local input and commercial investment is the key to what I believe is an important aspect of Cable Television, at least in Glasgow. If subscribers feel involved with us, if they feel we reflect their interests and if we provide a service which cannot be accessed any other way, then these subscribers will stay on the system. But everyone knows that local programming costs a fortune and anyway how can you demonstrate in figures on the balance sheet its effectiveness. And just look around you at the experience so far, I mean Aberdeen closed it down, Swindon closed it down, most new stations have never even started it - who's next for the axe?

I think there are a number of reasons that local programming has a poor reputation in cable. First, because in the early days in some stations it <u>was</u> too expensive. Second, because its too easy a target in the accounts - when times are rough its a lump that can be easily excised, without apparently affecting the body of mainstream business.

Finally I think that we have failed to demonstrate the commercial advantages of a local production capacity - something I'd like to put right now.

At Clyde Cablevision I don't see local programming in isolation, as an add-on extra, as an expensive left-over from the franchise application - included in the hope that worthy aims would help us win. I see local production as a partnership, a partnership with sponsors and advertisers, an integral part of the company's business, and most importantly a revenue earner.

So what are we doing at Clyde Cablevision ?

Well, first we're producing news - daily news headlines and a twice-weekly news magazine. We produce news because it provides a backbone. It provides us with set points in the day onto which adverts, promos and other programming can be added. It gives us a constant daily contact with our subscribers through which we can address them, their interests and their community. We have no expensive contracts with news suppliers and we only have four people exclusively engaged on local programme production and even one of them is the maintenance engineer on both Transmission and Production because of course we've got no costly maintenance contracts with dealers. They source write and present the news in the studio. They record and edit six to ten stories shot on location each week and they make an omnibus news programme that goes out on Saturdays and Sundays.

We cover only Glasgow news. More than 90% of the stories are actually sourced in our franchise area. That's one of the luxuries of cable - we know exactly who our audience are and we can target material accordingly. There's no point in us covering the same material as BBC Scotland or Scottish Television - they have to satisfy an audience made up of disparate communities, urban and rural with widely differing characteristics and interests - we don't. They also have vast resources and capital to do so - we don't.

... The response is tremendous. Local groups and individuals feel we care about issues that affect their lives, they enjoy watching television that shows their street, their shopping centre, their neighbours. Organisations get television coverage of their events (and if they want a copy we charge a fee).

... We have built up a network of relationships with Strathclyde Regional Council, the District Councils, Manpower Services Commission, Education Authorities, Health Boards and local businesses, all of whom command extremely large

budgets. All these organisations have experienced the response generated by having a story carried on Clyde Cable. This demonstrable proof has encouraged approaches from these organisations for airtime.

... Finally, I found it interesting to hear on another platform a representative of the advertising agency Leo Burnett expressing the view that the fact that the Financial Times makes more money from advertising than any other newspaper should tell us something about the importance of the target audience for advertising.

I submit that this is one of the strengths and attractions of the cable audience for advertising.

... I think that for the past two years we have been fostering the community to which we broadcast: and that advertisers, commercial and corporate bodies are at the point of recognising what we have to offer.

Footnote: In 1989, fifteen months after this address, Joyce Taylor left Clyde to join Discovery Channel and then moved on to her present role as CEO for United Artists Programming. As funding became harder the Clyde Cablevision system ceased expansion and subscriber numbers froze at below 6000. It placed its production unit on a commercial revenue earning footing, with some limited success. The decision was finally taken to cut local production in February 1993 but the system has continued providing a limited local service.

CHAPTER 7

THE ORGANISATION OF LOCAL TELEVISION

Models, arguments and case studies

A number of truths became evident as I began to develop this book. The first is that the US model of local television has little validity here in the UK. The fact is that the social, cultural and regulatory environments are fundamentally different. To approach UK local television from an exclusively north American viewpoint is a mistake. But to fail to recognise that there is a formula which can be learnt from this experience is equally a mistake.

Bob Garratt's excellent book *The Learning Organisation* has an intriguing chapter on changing behaviour by tackling the subject from the approach of "moving from, either ... or, to, both ... and".

Directors achieve this important management stance by valuing differences and becoming comfortable with them; by learning to value the asking of discriminating questions of experts rather than valuing depth of knowledge for its own sake; and by learning to value the skills and attitudes needed to set managerial questions in their broader economic, social and political contexts.

The outcome, he proposes, is that the most creative use of all parties' abilities is achieved. Failure to approach matters this way sets up a series of confrontational lose/win scenarios.

We can identify such situations in a number of UK cable systems. It is not just a matter of conflicting management styles. A starting position based on "them and us" creates a position where power and arrogance appear to go hand-in-hand and which, because it is blind to the detailed knowledge of native managers forces them through a cycle of self doubt, undervaluation, and finally despair. They either get out or are willing to leave to the ultimate detriment of the organisation.

One crucial issue in local television has been the number of US consultants who have recommended US patterns of programming and management. Why should such approaches be more successful in cable than in broadcasting, where the uniquely British approach has produced the most highly valued television in the world ? The only available answer seems to be, because some owners of cable systems want to look at it that way.

It appears that none of these consultants (and there have been quite a number) have covered any of the ground in this book, or even attempted to understand how the UK population creates expectations of the media in its locality. The British have often been accused of looking too intently at the past but the local TV lessons that are available tell a fascinating tale of experimenting on a real public testbed, discovering what works and what does not, both organisationally and in terms of local public positioning.

The earlier chapters illustrate that local television has been through a number of iterations. The first phase was prior to the 1984 Cable and Broadcasting Act, the second phase started with the first broadband franchises in the mid 1980s, and the third phase was the activity in the early years of the 1990s. There is every reason to believe that we are now entering a new phase which will be more enduring than any of the earlier ones. There are now more cable customers capable of receiving multi-channel television than ever before and the subscriber counts on several systems now exceed even the more pessimistic benchmarks offered in franchise applications for the start-up of local services.

This chapter is intended to help clarify understanding about the key organisational issues of local television and to demonstrate through simple case studies how these issues have been tackled by different companies.

Organisational Models

Since the beginning of the 1990s the leading debate in local television has been the matter of volunteers. There are those who maintain that quality is the key issue and that this cannot be delivered by volunteers; the debate being with those who believe (or have decided) that the main cost in programme making is the 'people' cost and the only way of achieving a meaningful local television service economically is to encourage a volunteer structure.

This debate is largely sterile. Both sides of the argument tend to agree that the total spectrum of local programming is likely to be a 'mixed economy', involving acquired, volunteer, commissioned and staff produced and facilitated programming. The mix will also include swopped programming, as channels acquire programme assets with a long life. Seeing what other local communities are doing will be another unique aspect in a schedule with an already unique flavour. The route taken to reach this ideal is what the debate should be about.

The history of local television in the UK reveals two leading models of organisation and a number of variants of them. Success however is a relative matter and whilst all the models may be regarded as successful against selective criteria none can be stated to be successful by any accounting criteria which might be applied by the accountants in a cable operating company. Neither, as we have seen, has there been substantial longevity backed by proper evaluation of the work undertaken.

With the stark exception of the early Greenwich and Swindon, the Clyde of the middle years, and the continuing example of Coventry, no local channel has had a life much longer than three years. The cable operator accountants' requirement that local television should be seen to cover its costs, despite the fact that there is no proof anywhere in the world that such a strand of programming can be profitable, is essentially destructive. These channels survived because their managements have commitment and vision. Any channel started from the enthusiasm of a visionary manager is vulnerable as soon as he (or she) moves to pastures new.

To encourage today's local channel enthusiasts there is growing evidence that the advertising world is fragmenting alongside the growth of new niche channels. With a number of channels offering 'ad avails' in which a cable operator can insert his own advertising he is encouraged to move into the local market in a new way.

If local advertising on TV can make a local operation financially viable and if the hidden benefits of local programming can be realised (ie. reduced 'churn' and better penetration) the arguments in favour of a local channel being subsidised through this new revenue route become almost undeniable. A failure to do so will open cable operators up to accusations of milking the local economy and putting little back in. However unjustified that may seem in view of the investments cable companies are making, local people will be encouraged look at this issue in the light of franchise promises.

This will become more apparent if the commercial base of other local media (eg. the newspapers) becomes threatened by local cable advertising. There are already political moves underway in the local newspaper industry (which employs over 250,000 people) to protect their markets. Cable may not be immune to legal actions under monopoly or fair trading laws, and any such action against it could create a damaging groundswell of opinion - locally or nationally.

What is needed is a working model to encourage operators (or third party channel providers) to venture out. The two models that can be readily identified from the experience of the last eight years are the broadcast model and the Canadian model. The principle characteristics are as follows.

Broadcast Model

This model is best defined by the commissioning of programming from independent producers and the provision of a core of regular strands of programmes either by staff or by commissioned producers. The mix of own produced programming and programmes from external sources is scheduled by a channel manager or executive producer (by whatever name) working for the cable operator, and acting as the local channel's senior executive (in practice if not in name) as is required by the ITC channel licence.

Programming is produced in-house or in regular commissioned programme strands. It is likely to be topical, studio-based and represent the maximum air-time at minimum cost. Such programming requires a significant staff, and it is this cost which needs to be represented by the maximum air-time. In this model the channel manager also coordinates the activity of local voluntary or unpaid student programming contributors, and buys programming as and when there is need and opportunity.

Some of the best representatives of this model are in the past. The first examples to start-up were short-lived, Swindon in 1984, Aberdeen in 1985, and the original Coventry approach. The organisational core in Clyde relied on a broadcaster structure though there were some significant differences from a broadcaster in its approach.

Swindon, Aberdeen and Coventry all recruited editorial, production and technical teams with the main objective of running a daily news magazine programme.

The Croydon approach even in the earliest days was a modern modification of this model. Although changing hands several times since it connected its first customer in 1985 the London South franchise block is now nearly four years into the life of its daily news magazine programme. The production was originally commissioned from an associated company, Crystalvision, though it is now managed in-house.

One key feature of the broadcast model is the relatively high staff head count. Put together with the inevitable capital investment in equipment this offers a high start-up cost and continuing high on-costs, both of which are easy to question if the operation's performance indicators are purely financial.

The Variant

The latest channel to start-up which draws its main influence from this broadcast model is Birmingham Cable's Channel 021. The approach differs by taking examples of good practice from many of the other lessons of local television already outlined - the key lesson being the recognition of the need for a low overhead.

Channel manager Judith Mackay acquires programming which already exists in the local market, locally relevant programmes from the national market, and commissions programmes from local production units, but makes none with in-house effort. The commissions are on a non-exclusive basis and the programmer is expected to make his profits from sales in other markets. The channel does source its own text/graphics, which carries news supplied by the Birmingham Post and Mail group - the leading regional newspaper, financial information (from a local stockbroker) and a carefully selected range of local and community information.

The Canadian Model

The main feature of the Canadian model is its heavy reliance on volunteers. A small staff team, usually a manager/producer supported by technical staff and a volunteer co-ordinator, create a structure which trains volunteers, provides editorial guidance and control, decides which programmes offer the most appropriate balance for their schedule, and then helps co-ordinate volunteer activity so that programmes get made.

There are likely to be long running strands of regular programming (Leicester is a notable example). One feature which is consistent to both Canadian systems and their UK emulators is the difficulty of getting volunteer producers to finish their programmes in time for a scheduled transmission, even when the schedule is planned well in advance.

This model is also evidenced by the (now defunct) Channel 7 on East Lancashire Cablevision (originally a Maclean Hunter franchise), Leicester Communications (whose influence came from the Nova Scotia based Fundy Cable), and Swindon (now a Telecential franchise). Windsor/Middlesex Cable (managed by General Cable) began as a British modification of the Canadian model - created by Sally Allen, who was the first channel manager of the East Lancs system.

East Lancashire's Channel 7 was a vision of Ed Madden the system's managing director. Ed was the senior vice-president of Maclean Hunter who set up and ran their UK company until the decision to sell to pull out in 1993. Ed is a committed programmer, and established local channel operations in Canadian new build franchises from their very earliest days. His commitment to that approach during the build of East Lancs, the only Maclean Hunter franchise to start in the UK, resulted in designing and building a 3-camera OB truck and an extensive editing and transmission suite. He closely followed the long established organisation of local programming adopted in the both the US and Canadian franchises owned by Maclean Hunter.

The Leicester system started programming when the network was about a year old. The system was languishing with a poor penetration and faltering marketing in the wake of a stalled build programme. Local programming began with volunteers. Like the original East Lancs Channel 7 its success is as much due to the energy and commitment of its manager, Jenny Ludlam, as to any other factor.

Sally Allen's experience in setting up East Lancs Channel 7 over 18 months in Blackburn made her uniquely attractive when the management in Windsor decided that local TV was an appropriate way to tackle their marketing problems. Her approach in developing WMTV called on volunteers but also drew on the broadcasting model as the budget allowed her to employ freelances and thus make available a wider range of professional programme making skills than an exclusively volunteer supported channel operation can usually find. The compensating factor was to take advertisements and the commitment to raise revenue.

At the end of 1993 WMTV had drawn in its ambitions and evolved into a broadcast model of largely in-house production.

Once CUC Cablevision had acquired Swindon in 1989 it's sole staff producer, Paul Langcaster, ran a daily schedule calling heavily on volunteer effort. Like many of its Canadian colleagues CUC faced financial problems during the early 1990s and although completion of the Swindon network took place on schedule it is only as 1994 dawned that programming could really begin to expand in line with the Canadian influence.

A further example already in existence is the all-volunteer charity, Parallel Pictures, supplying Telecential's Hemel Hempstead based franchise. Parallel Pictures attracts funding from various local grant bodies to enable it to supply programming. Its founder, local journalist Keith Jeffries, became the staff producer in early 1994 and through this dual role took effective control of the channel.

The Variant

Cable London's local channel was set up in 1990 purely as a text/graphics channel and with the objective of moving into TV programming during the following year with volunteers. The plan was that only a single member of staff would be needed initially as the unique nature of the local community meant that a high degree of editorial and production skills were available. This was borne out in practice and for two years the quality of programming produced by volunteers with no staff help at all exceeded the expectations of those involved who had long experience of programme making.

The Cable London channel was closed by a new management in 1993. Some concern existed about the ability of the company to retain full editorial control with the number of volunteers that then existed. The real successes in line with franchise promises and the fact that costs to the cable company were very small, has led to a negotiated arrangement between the company, the core of the volunteer group and a local media organisation (LVA) to continue to provide facilities and airtime to programme makers.

Cable London's organisational model was a variant, calling more on the Canadian model than the broadcast but as in the case of Birmingham, not wholly applicable to either.

Lessons

It is likely that many new channels will take some elements from both these core models. One lesson from the Cable London experience is that however well editorialised programming may be it has to be seen to be controlled by the licensee. When the nominated manager of the licensee does not have day-to-day control of every aspect of volunteer programming there will be an insecure feeling.

This is not a problem when a third party channel is concerned. Such channels have their own licence and are directly responsible to the ITC for implementation of the codes. The cable company can regard them just like any other channel carried (although they might want assurances that no programme is going to criticise them).

Third Party Channels

For third party groups or companies seeking to run local television channels the approach is somewhat different to the one that a cable operator will take. Balancing the need to generate large quantities of quality programming with the need to control costs is more difficult for an independent local company because it does not have a subscriber penetration imperative or the secondary benefits that a cable company can use to justify costs.

We discuss likely partners and third parties in Chapter 13 looking at the business of local television. I believe that third party channels are viable and if based in the community, with the commitment of a strong volunteer group, they can be successful both commercially and in terms of viewer satisfaction. The commercial equation will eventually be determined by the ability of such channels to sell airtime in the local market although programme and 'slot' sponsorship is also likely to play a significant role.

Third party programme providers offering programming on a nil net cost basis can also be found to contribute to operator managed channels.

The benefits for a cable operator are those of fostering a community operation at minimal cost and of gaining the warm recognition that comes through the 'feel good' factor. The benefits are widely believed to include reduced 'churn' (though there is as yet no empirical proof), an easier relationship with local authorities, subscriber loyalty helping the conversion of doubters and building better quality penetration cost efficiently, and actively cross-selling other products such as business telephony - a market which will get more and more competitive as BT adapts to the new market conditions.

What is in it for the third party becomes a matter for negotiation. A per subscriber fee will clearly be sought, though this may be countered by an advertising levy for the cable operator. Advertising is clearly a necessity and the following models illustrate some other early approaches.

Open Campus Television OCTV

The key exemplar third party channel with a non-commercial background is Blackburn College's Open Campus TV in the East Lancashire franchise.

OCTV offers a schedule of educational programming which comes from three sources;

- acquired programming - the library of the Central Office of Information (COI) plays a key role
- the College's own production effort - which also attracts external funding for specific educational projects
- the use of college facilities by students and local schools.

The channel only runs during term-time. It has access to the College production facilities which include a studio, and has editing and transmission facilities under its own control. The funding for the operation is through the College's own budget, and their initiative has been recognised and well supported by education and training funding agencies keen to examine the potential this experience will demonstrate.

Following the takeover of the East Lancs cable system by NYNEX Cablecomms in 1993 they have affirmed their commitment to carry OCTV and to make it available in due course in their other franchises in north-west England.

The MultiCultural Channel

In 1992 Videotron in London launched a weekend schedule of five hours of television, taking advantage of the fact that the cable-exclusive Parliamentary channel would be off-air over the weekend. Five programme blocks were offered to five ethnic/language groups who could schedule what they wished into their allocated time, subject only to the requirements that the programming came at no cost, and that each provider acquired his own ITC Programme Service Licence. This way Videotron did not have to face the problem of exercising editorial control over non-English language material.

The five groups included Channel A, an Asian programme service based in Leicester and run by the Visavadia brothers from their home, and Hellenic Television, set up originally to provide a Greek language premium tier channel in North London. Channel A were succeeding in funding their operation through advertising and sponsorship.

Hellenic Television

Dr Chris Fellas, a broadcasting and satellite transmission systems engineer living in North London contacted Cable London in 1986 about a Greek language channel for the area, which has more than 25,000 Greek speaking households. His proposals were included in Cable London's later franchise applications. In late 1991 the channel began transmissions as a premium service, costing those who wished to take it £5.95 a month.

Initially putting out 4 hours of taped programming from Greece and Cyprus plus inserts of local news and events coverage, it soon installed its own downlink and made the necessary agreements to re-transmit Greek television, allowing it to grow to a 16 hour daily schedule. The first advertiser was a local Greek greengrocer, but airtime sales have been difficult as the reach is so small. In mid 1993 there were 800 subscribers in Haringey and Enfield.

Hellenic remains the only local television channel ever transmitted as a specific pay channel.

CASE STUDY
CHANNEL 7 EAST LANCASHIRE

Channel 7 East Lancashire was shutdown by new owners NYNEX Cablecomms in October 1993. The assets and remaining staff of the channel have been loaned to University College Salford and will provide the core of a new regional television operation covering all NYNEX's north-western franchises.

In December 1992 Channel 7 held a charity telethon. The following report appeared in 'Inside Cable' in April 1993.

The director was a fireman, the main presenter a local DJ, another presenter was a videotape operator, and one of the cameramen had been a student this time last year. The graphics operator was still an FE student - and they went on air for a four hour charity 'telethon'.

The studio was normally the office, and the 'outside' location was the building foyer. Two cameras in each location (no spares) and eight 'redheads' for lighting. The control room was an OB truck parked three floors down, and the starting time was ... just a little flexible.

Not something from Edward Lear, but a programme from 'Cable 7', probably the most active local TV channel in UK television.

When presenter Russell Mclean smiled his way onto the TV screens of Blackburn and Darwen just before 6.45 on a Tuesday evening in December he carried wishes expressed by the crossed fingers of the staff of East Lancashire Cablevision (many of who were operating phone lines to help with the programme) as well as those of 14 other volunteers.

His personable introduction would have done credit to a Wogan, and even if four cameras, two VTRS and a Commodore Amiga graphics computer compare unfavourably with the resources available to 'Children in Need' (transmitted to 15+ million viewers), Russ and the East Lancs team, whose efforts were viewed in fewer than 5000 homes, made just as important a contribution to their community, and every bit as professionally.

'Cable 7' started its service in 1990 and has grown to transmit a total of 16 hours every week - with new programmes in the schedule every Monday and Tuesday. The range of programming is wide - local sport and events, discussion programmes on major topics, music, review of local issues, and documentary programmes looking at everything from diet to local history.

East Lancashire Cablevision's Managing Director, Ed Madden, is an enthusiast for local programming. In Canada he helped establish several local operations for Maclean Hunter, and when he came to England to start up their first UK franchise he set out to replicate that experience. A 3-camera outside broadcast truck was locally built to a basic Canadian design, and a team of staff and volunteers was recruited and trained.

Ex- journalist Milton Haworth took over as Local Channel Manager for Cable 7 earlier this year. He has a staff team of 3 to help him create and manage one of the most ambitious, flexible, and under resourced, TV production environments in the UK.

The staff - Claire Haddock co-ordinates and trains the volunteers; Jo Siddiqui works as producer/director; Glen Hunt edits, operates graphics and plays-out the channel; and Chris Knagg comes in to work as engineer on productions (as well as being the company's Construction Supervisor) . Like the volunteers, all are multi-skilled and are as likely to be floor-managing as working on camera as directing, or presenting.

Multi-skilling is a key feature of cable programming, and well evidenced by the Cable 7 telethon, 'Christmas Cracker'. One VTR insert featured a review of videos available this Christmas, produced and presented by Ashley Heath. At the end of the insert, cut back to Russell in the studio, and who should be cameraman for that shot but the same Ashley Heath ! They don't think twice about it.

Thought of a volunteer production environment will send many broadcasters into paroxysms of horror. Yet it is clear from the Cable 7 experience that volunteers can deliver programming which is highly viewable, and has a unique attraction through its spontaneity and preparedness to challenge the conventions that professionals bring to the technology. And that is not to suggest that their work is amateurish or slapdash - the standards that volunteers set themselves are as demanding as those they see on other TV channels, and they are as able to assess their own competence as any other viewer.

'Christmas Cracker' was researched and produced by Russell Mclean and Neil Jordison - both volunteers. The directors were Jo Siddiqui and Nick Isherwood (Station Community Officer at Accrington Fire Station). Claire and Glen shared the presenting with Russell and between them they worked through more than 70 guests in 4 hours - guests who sang, danced, were quizzed, talked about Aids and local radio, reflected on the cinema, talked about the work of the Lions (in support of whom the programme was made), offered advice from British Gas, and were wished Happy Birthday.

Pre-recorded inserts offered the occasional breather, and a view of Blackburn's Christmas lights, the channel's out-takes and selected highlights from the year's programming, while the opportunity was taken to re-set as the piles of tins and packets of food for the elderly and poor, donated by viewers, grew. Every gift attracted a draw ticket for a range of prizes donated by local companies - the most sought after of which was a football signed by the whole Blackburn Rovers team - including Kenny Dalglish.

As the pubs closed the pace grew frenetic. Russell quizzed caller after caller - "which king was known as the Hammer of the Scots ?" ... "who invented paper money ?" .... "which actor featured in ... ?" etc. The office loudspeaker telephone (with microphone attached) coped well with its unusual function.

Just over four hours after setting off on the marathon programme, the Channel Manager (also floor managing), decided to call for the closing titles. The prizes were running out, the production crew was exhausted and, even if the presenters' adrenalin was still coping, the production plan had worked well and as 11pm loomed closer they had run out of guests. Mr Henderson and his children collected their football prize (they donated a box with 45 cans and packets gathered from neighbours) and the phone stopped ringing.

Channel Manager Milton Haworth explained, “We've got to get clear tonight; tomorrow we've another long evening - in town - but at least we're not going out live,” he added.

As the crew began the de-rig Managing Director Ed Madden quietly started putting the office back together. “It has to be back to normal for the morning,” he said, “we have to run the business in here too.”

CASE STUDY
CHANNEL 7 LEICESTER

The Leicester franchise is owned by a consortium comprising John Laing plc, Fundy Cable and SaskTel - the majority shareholder being Sasktel. The decision to set-up Channel 7 was taken when the franchise was at a standstill in early 1992. Jenny Ludlam, well known in the local community, was recruited as manager of the channel and after a training trip to Canada she started the channel three months later. The following report appeared in 'Inside Cable' in June 1993.

Simon Kimber's opening link to camera was delivered with the assurance of one with long practice in TV reporting. Kiran on camera and Jason on sound were indistinguishable from the nearby BBC crew.

Back in the office the manager of the TV channel was sorting out the details of the schedule for the last week in the month, and checking with another of her producers that his programme would be ready in time.

In the edit suite the evening transmission master tape was being finalised and graphics prepared on a computer.

An everyday story of TV folk, but one that the people involved would not have countenanced just a year ago.

Leicester's Community Channel launched in July 1992 with a simple text service and began transmitting video programming in September. Manager Jenny Ludlam had no experience of TV when she was appointed in March 1992 and admits to approaching the job with some trepidation. She now manages a schedule which transmits up to 8 hours of new local material each week in a schedule which changes three times during the week. Most programmes are transmitted 6 times.

It is an example of local TV in action which defies any broadcaster's doubts about the ability of a low-cost volunteer based service to sustain a TV channel.

The resources available are minimal - two single camera S-VHS location units (one a camcorder) supported by a

2-machine editing system and Amiga graphics. While these are the only technical facilities that are available as of right, a local authority media centre also contributes some effort. This slender resource highlights the importance of people. In addition to herself as manager and executive producer Jenny can call on a full-time engineer, Kiran Raval, two part-time operators, two students on work placement two days a week, and a small army of volunteers.

Decision-making about the channel is in the hands of Jenny and two 'co-ordinators' who are, in effect, also executive producers. Martin Mitchell is a non-conformist minister with extensive experience of local radio and his time spent on TV programming is part-funded through the local council of churches. Alan Wilson is the organiser of Fosse Community Studios, the local authority supported facility which offers open access to the community. Programme ideas are managed to completion by one or other of these three, supervising production and ensuring that the Programme Code and other regulations are observed by the volunteers.

Foremost amongst the volunteers are the 'do-ers', the ideas people, those who have undergone training to enable them to bring programme ideas to life, and those who can devote time to helping in the office and with admin duties. A User Club is being formed so that these active volunteers can 'network' together and can express a common voice.

At a lower and less involved level, but none the less crucial for local accountability, is the Cable Forum. Open to anyone in the Leicester community the Forum meets two or three times a year. People are able to offer ideas, comment on what has happened already, and use it as a route to becoming more active if they so wish.

Apart from the volunteers Jenny Ludlam has built links with the local council, local colleges and the universities. The student union of one is already producing its own programme of student oriented entertainment - the catchy title "LUST" referring to Leicester University Student TV.

Other programmes cover the whole spectrum of local TV potential. Sport is attractive to viewers, relatively simple to produce, and inevitably accounts for a significant proportion of programme time. Other regulars include "Audrey's Allsorts" - featuring a lady of maturer years who goes out to find people to talk to and events to get involved in; local bus news - presented by the local managers; "Woodside Waifs" - from the RSPCA's local home; and the "Odd Show" - a youth entertainment programme which baffles adults but has acquired a cult following among teenage viewers. And all this in just nine months !

A WEEK'S SCHEDULE IN LEICESTER

Fri/Sat/Mon (Also at 9.00pm)

5.55 Cable 7 Shop Window
6.00 Crossview - the local church scene
6.30 Woodside Waifs
6.45 Bob Hammond's Red Cross Special - looking at accident simulation
7.00 Filbert Focus with Simon Kimber
7.30 Audrey's Allsorts - Audrey goes to the Rutland Water Butterfly Farm
8.00 Easter Egg Run - motorcyclists, eggs and children.

Sunday

3.00 Basketball - Leicester Riders v. Birmingham
4.30 Football - Cosby United v. GEC SSC

Tues/Weds/Thurs (Also at 9.00pm)

6.15 Cable 7 Shop Window
6.20 Junior Football Results with Dave Hardy
6.30 Midland Fox News
6.40 The Odd Show - more fun from Guy Harris
7.00 Leicester and District Football Results
7.10 Rugby League - Leicester v. Peterborough

PORTRAIT OF A LEICESTER VOLUNTEER

Simon Kimber is a sales representative for sophisticated medical electronics products. A lifelong football 'nut' and devoted Leicester City fan he presents "Filbert Focus" every Friday.

His approach to his reporting is as a fan. He is adamant that what makes his reporting different from that of the traditional TV companies is that he talks to the manager and the stars as someone who cheers them on from the terraces each week. This is not just talk. In his questions to players and manager Bryan Little, Simon refers to his subjective feelings during the last match. It gets the interviewees responding in a more intimate way. He admits to consciously retaining a sense of awe when he talks to the stars, believing that it helps him retain his identity as a fan.

About his commitment can be little doubt. He fits his reporting into lunch times and immediately after we met him was off to see a customer in Birmingham.

CASE STUDY
COLT, COVENTRY

It would be possible to write an entire book about the experiences of the Coventry local channel over the years since it was launched in 1985. The first iteration was a broadcast model, the second an adapted Canadian model and the latest is a midway house.

One of the key programme providers is Coventry's Hereward College, the National FE College for the Disabled. Hereward's Principal, Rees Williams, spotted early on that technology, the arts and the media offered disabled people an unrivalled opportunity for employment and he set out to create a range of courses which could fit them for such work.

As Coventry Cable changed its emphasis in 1988/89 he saw the opportunity to become a programme provider and now the College produces more than 9 hours a month of TV for the local channel, calling on its students, both able and disabled, to help in the production work.

Martin Potts works as full-time producer and production co-ordinator for the College. A major area of activity is local sports, where basketball is proving a big attraction. Once the cable channel began covering the matches the actual attendance at the games also rose !

Coventry Cable has two edit systems, three location camera units and various supporting equipment, as well as a studio. The College is free to borrow these as available, and also uses its own cameras, editing, graphics and studio space as required.

The COLT schedule offers up to four hours each evening of locally sourced programming, much of it repeated several times during the week. There are no complaints about the repeat frequency, over the years local people have come to recognise that the channel is different - offering many opportunities to see a programme during its short 'shelf life'.

One of the advantages of a long standing recognition in the community (however small the actual subscriber numbers connected) is that innovative ideas are tackled without fear.

A recent example is the collaboration on a sponsored programme. Local brewery M&B has joined forces with the Coventry Evening Telegraph and Coventry Cable to sponsor a quiz - "The Knowledge". The quiz is highly competitive in the M&B pubs in the city, and a team prize of £500 is offered to the highest score each week. Home viewers can join in with a form printed on the day of transmission in the Evening Telegraph - though they cannot compete for the £500 pub prize. The quiz includes general knowledge questions, but also has pictorial questions drawing on city sights from unusual angles (eg. rooftops) and historical photos from the newspaper's files.

It has been an outstanding success for all three sponsors and before the first series was completed in autumn 1993 the second series had been commissioned. With the system penetration at barely 10% this high profile programme, receiving full commercial sponsorship and carrying advertising in conscious 'comfort' breaks scheduled for the benefit of the pub audience, does offer more than just a crumb of comfort for those who believe that local television can be financially viable through programming which brings in direct revenue.

CASE STUDY
CLTV CABLE LONDON

Cable London's channel started with a simple text system based on a BBC microcomputer in 1987. This was replaced in mid 1990 with a graphics based system from Picture Applications. A small investment was made in Hi8 production and editing kit and plans were laid to develop community based, volunteer produced output from mid/late 1991.

A mature film/TV student, Shan Stephens, joined the volunteer group in 1991. One assignment on her degree course was a dissertation on an aspect of the media - she chose a study on CLTV and in particular aspects of its reliance on the volunteer group.

The following are extracts from her dissertation, starting at a monthly volunteer group meeting.

Over 100 people drifted in during the course of the evening. The age group averaged around the 20s and 30s although there were a small proportion of both under 20s and over 40s. Ethnic members were healthily in evidence. Cloning in clothing style was minimal; it went through a range of trends from Cosmopolitan to student indifference in casualness...

...Proposals for programmes on the channel were put to the floor. Each Producer of a proposed programme was invited to the front to give a brief presentation of their idea. They were then at the mercy of questions from the audience. This took place somewhat hectically and on the whole humorously. Carole Enahoro (the Channel Manager) then moved proceedings on to review programmes. A short clip of the programme selected by the producer was shown. A potential new member brought the show-reel he had put together with a view to becoming a television presenter. It was extremely professional and nattily done. DEFII would have been proud. A few minutes of a documentary showing a job training venture for chefs, mechanics and carpenters in Camden was shown. A very unusual and witty view of Kings Cross station was previewed,

followed by a clip of a 30 minute love story drama. Each of these polished programmes was voted through after a few minutes of criticism, questions and compliments from the floor. Some of the women thought the drama veered dangerously near to being sexist, but it was passed as dramas at present were far and between. ...

... The amount of time and energy which people were prepared to put into the making programmes is quite extraordinary. It was clear in my discussions with club members that the vast majority of them saw CLTV as providing a format and outlet for their creativity and ideas which was not available elsewhere.

It is this opportunity to broadcast which Carole sees as creating the dynamic to the energy in the club. Overall, she is very positive about the group. "They energise each other and I believe enthusiasm does travel". Carole believes the standard of programming to be high as the makers have to defend their product publicly. "It is an unusual forum of debate for TV - people do not usually have the opportunity to publicly challenge programme makers."

"The main problems seem to be with attitude. Because a lot of what the club offers seems to be free people take a rather lax attitude. They think I am there just to serve their every whim." She says there are quite a lot of rude members. Initially she thought it was personal but once she involved other organisations with the club she discovered they had similar problems with the same characters.

Whilst Cable London took the decision to extract itself from local programme production in 1993, this varied and energetic band of committed people has reformed with support from Cable London and London Video Access (LVA). It will be producing local programmes from summer 1994 onwards and Cable London will be showing them.

CHAPTER 8

THE PRACTICALITIES

An incremental approach to developing a local TV channel

Running a local television channel is a matter of energy and commitment. There are no such things as early evenings or quiet weekends. A channel manager must be energetic, visionary, have his (or her) feet well planted on the ground, and be both determined and eternally tactful even under great pressure from every direction imaginable.

The reality is that when the schedule is written, the programmes delivered and cued up ready for transmission, the graphics computer programmed and all its information up-to-date, the shooting equipment in action on a planned programme, the (few) staff beneficially employed and the editing equipment busy, there is still work to do. There may be volunteers to organise or recruit, community groups who need a visit, encouragement or help - and management which needs reassuring that the year's budget is still being observed. That may all be before the salesmen are due to report on their successes (or difficulties for the month) and the credit control clerk comes in worrying about the payment record of the biggest advertising customer. Then there is programme acquisition, relationships with collaborating production companies, other cable operators ... the list of possible pressure points is endless.

For most cable local channel operations this list is far from being a reality even though there are staff who can conceive that an operation with these demands will develop and believe that reaching this level of activity is achievable in their company. Until then a more prosaic reality has to be built.

Starting Out

Local television channels usually begin with a simple photo-videotext (multimedia graphics) computer. A single computer and some graphics design skills are all that is needed to get on air. This is a very thin resource as many cable companies have already found out. Illness, technical breakdown, software faults - all can take the channel off air and erode its credibility because a topical service cannot be sustained.

During 1987-89 I maintained a simple channel within the Cable Camden network on this basis. We kept a 'carousel' of 40-50 text frames running on a BBC Microcomputer based on 3 hours work a week (serving 35 customers). It was a salutary learning experience about the impact that even such a simple local channel could have. Spelling mistakes, out-of-date information, and breakdowns all attracted telephone calls, even from such a small subscriber base.

The BBC computer was very low-cost and could not handle graphics. Today a good graphics computer and software to create attractive screens and schedule them can cost less than £10,000. A designer is needed, information gathering and editing is required, and there is the need for management time to be allocated. The designer could be a recent design graduate in his or her first job but do not expect someone with this experience to also be an effective information gatherer or competent editor.

There have been attempts to run this structure with the information gatherer undertaking the graphics. It does not work. Computer graphics and text may be easy to learn to produce but it is even easier to make the output look awful. In one unnamed UK franchise I saw text screens from its computer which averaged four different fonts in six colours and a variety of different sizes. The screens looked a mess, but because they were local information the company was proud of them. They were not wrong to be proud that they were carrying local information, but it was extraordinary that marketing people who could be pernickety about the detail and design of their printed literature seemed to be blind to the absence of similar standards in their output on the screen.

Even the two person organisational structure is too frail to sustain a topical service for long. Staff need holidays and in a small operation the initial creative energy soon dissipates in the strain of sustaining the service. The moment one member of staff leaves (as they surely will) the benefit of the built-up experience disappears and a major re-training exercise becomes necessary.

Unfortunately a practicable level of effort and resource is also more costly. So the solution to the financial dilemma is usually to seek revenue from advertising. Can the information gatherer also sell advertising ? There is every reason to believe that the holder of this dual role cannot succeed. The skills are so different and the potential for conflicts both of time and interest is enormous.

Failure to recognise theses stresses means that a sub-standard service is provided. By creating a situation where all the worst stereotypes of text/graphics systems and of local television standards are confirmed, even when subscriber numbers increase there is little management incentive to improve matters. The dilemma of a budget justification to provide staff cover, back up equipment and, a dedicated salesman for what is regarded as a sub-standard service is one that a number of people have had to face. Building the next steps in such a situation is equally difficult.

My advice to anyone entering at this simplest level of local television service is to:

- define medium/longer term objectives and have them agreed at a senior level
- define realistic and achievable performance standards
- define the roles necessary for a quality service
- appoint suitably qualified people to each role - even if in the first instance they are part-time
- support the team as a production department, not as a clerical off-shoot of marketing.

Correlating the effort to the numbers of subscribers served is an initial approach which a number of text/graphic channels have attempted. Where appropriate subscriber levels for particular levels of resourcing lie is a matter of debate and there is little evidence that this approach is actually valid.

A simple truth about television services is that it costs no more to produce a programme for 1 million viewers than for 1 viewer. I would maintain that starting a service properly resourced and with a clear vision of how it will grow should only be subscriber number dependent at critical resource cost points in its predictable life. Following the creation of an initial text/graphics service the next such stage is the introduction of 'real' television programming.

The resourcing of this move is likely to be significant. The best indication that we can draw from market experience in the UK to date suggests that a viable subscriber number which can generate advertising revenue sufficient to support the introduction of programming, is probably in the region of 35-40,000 households. But this is only one of a cable operator's considerations.

A cable operator can decide to budget a local channel independently of its direct revenue. The value to the channel's budget that an added pound of income offers is the same regardless of the number of people that receive the programmes. Thus a company commitment to allocate funding which grows in line with subscriber growth allows a channel manager to have guaranteed revenue growth with which to plan. Growing revenue will allow him to demonstrate a disproportionately greater improvement in quantity and quality of programming.

Into TV Programming

The big step is the move to production and transmission of a local schedule of moving image television programming. There may be an intermediate stage where acquired programming is scheduled. The first questions to be answered are:

- what is all this programming to be, whether it is acquired or locally produced?
- who is to provide it, and at what cost?
- how will it be paid for?

Virtually every piece of research on local programming shows that local news comes the highest on the audience 'want' list. Yet, as we have already seen, news programming is one of the most resource intensive (and thus expensive) strands of programming which can be created. In addition news is largely unrepeatable following the day it is produced so the effort expended has no shelf-life. It also requires the maintenance of on-going resources to handle live, or as-live output. Thus a second list of questions arises:

- is local news a viable option1?
- are there any local alliances which can be fostered to provide it at a low cost?
- with or without news, how do we start planning a full schedule for a local channel?

Identifying the channel's development strategy is critical. Is the channel to be a commercial operation or is it to be run by the cable company as a feature of its community relationship, or his marketing presence (or a combination of all these activities)? Many start-up channels will be funded by the cable operator, probably through the marketing budget, but even if the plans are for an independent 'third party' channel the following basic arguments still apply.

Within the models of organisation we have already examined there are plenty of ideas about the kinds of structures which could be adopted. Judging how well they have worked is not easy in these early days when commercial activity has been very limited, especially as market research into the impact of such channels is more noticeable by its absence than its presence.

The starting point is direction from a Board on the channel's performance targets. If the channel is to be a major plank of the company's selling strategy a news programme is likely to be essential. However this does not have to be a live, hard news, heavily illustrated programme - Croydon's 'Cable Today' is a soft news magazine with topical forward-looking items, and a short hard news segment supplied by by the local paper and recorded late in the day. It is a successful formula.

Daily programming of this kind requires either an in-house team or a local production contractor who is able to deliver the product reliably day in, day out. Such a commitment is not a trivial issue and I would want to see proof of an employee's or supplier's experience before even discussing the possibilities with him. But even with two repeats an evening such a programme does not on its own, make a schedule. And as it is likely to be a five-day a week operation (for staffing reasons) something different is required on weekends. The total cost, even with the operating overhead cut to the bone, is not likely to be of the order of £400,000 per annum. Such sums understandably daunt even the biggest cable companies.

There have been, as yet, no emulators in the UK of the Croydon operation although TeleWest has suggested that others of its franchises may adopt a similar approach in due course.

Without news it is unlikely that a local channel can be a major element in the sales promotion of the cable company. In the view of most cable companies such a service will take more of a 'nice to do' or 'helping to build customer loyalty' theme.

If there is likely to be a medium to longer term ambition that the channel will become a profitable operation in its own right the news option will be regularly re-visited, until such time as a start-up is decided upon. Liaison with a local newspaper probably offers the most flexible route and could have the advantage that an extensive local ad sales force also becomes available. As we have seen the other obvious local news source, commercial radio, presents sufficient conflicting practices to make an alliance difficult.

Whatever the route a low-cost approach is likely to be the dominant theme.

Staffing

The inescapable issue in television programming is that production consumes manpower and uses relatively expensive equipment. The choices therefore lie between acquiring programmes which are already finished, or finding a route to producing a viable quantity of programmes within the budgets that are available. It is unavoidable that there will need to be some staff, and some investment in hardware - even if acquired programming is to be the dominant element in the schedule. Master tapes have to be assembled for transmission, trailers and station idents have to be inserted and there may also be advertising to be produced and inserted into programmes. Hence further questions can be asked.

- how large a staff team do you need?
- how many technical staff do you need?
- how much reliance can be placed on non-staff effort?

It is at this point that the merits, or otherwise of operating a volunteer structure have to be considered. Some will discount them immediately and look for a structure which accepts limited local production, commissioned or acquired within the available budgets - and hoping that the budgets will increase in time as the network grows.

The arguments in favour of a wide variety of locally produced volunteer programming are very persuasive. Above all it comes at a minimum cost. There is now ample proof from the USA, Canada and the UK that professional quality programming can be produced with volunteers playing key roles in the process. In the case of Cable London between 1991 to 1993 there was consistent success with programming left almost exclusively to volunteers.

The Windsor approach, using a mixture of staff, freelance specialists (especially directors) and volunteers has also been successful in delivering quality programming.

Ambitions for large numbers of hours of volunteer production are unlikely to be realised. There is no definitive measure but a simple rule of thumb based on UK experience to date suggests that somewhere between 8 and 12 hours a month is the most that can be expected per 100 volunteers. This goes down in direct proportion to the demand for increasing programme quality, and at the same time the training demands increase. And it is worth remembering that the more volunteers there are the more staff are needed to administer, motivate, provide quality technical support, and importantly to provide the necessary training.

Control and Access

Some resistance to local television has been apparent among US cable managers working in the UK. It seems to be based on experiences of access television in the USA. Rules based on the First Amendment to the Constitution require cable systems to offer free access to anyone with something to say. This right is exploited by all kinds of groups, including 'outing' gays and the Ku Klux Klan, sometimes in the most outrageous ways.

In the UK legislators and regulators have never been inhibited by the same fundamental level of legislation on 'free specch'. Public control of community television is more strictly applied. Every channel must be licensed, whether the cable operator or a third party body runs it. Editorial control of a local television channel rests with the licensee who is answerable to the regulators through the ITC Codes. This role is likely to be the day-to-day responsibility of a channel manager, even if the nominal licensee executive is someone more senior.

Editorial Policy

An Editorial Policy is an important statement which should be made public and be clearly understood by the staff and volunteers working for a channel. It would start with a statement of the channel's purpose designed to give focus to everyone's work and to give the background to decision-making when difficulties arise, as they surely will. For example:

The objective of (the local channel) is to provide a service of locally produced programmes covering events and matters of interest to the local community, and to schedule programmes of relevance and interest to local audiences even if produced elsewhere.

It may continue with clear guidance on content:

The target audience is the family. Presenters and producers will ensure that programmes and on-screen behaviour are suitable for the family audience. Whilst accepting a policy of equality of opportunity and of representation (the local channel) will not be bound by any perceptions of 'political correctness' in use of language and will encourage use of the vernacular and colloquial language as appropriate to the subject. In the event of any doubt advice should be sought from (named post).

It may also be decided that a statement about rights of 'access' and 'reply' may be appropriate.

There is no right of access to transmission time on (the local channel). However local people offering ideas or wishing to become involved in programme making are encouraged to participate and appropriate training is offered. In the event that a subject is dealt with in a programme and a local person directly affected feels that a reply would be relevant (the local channel) will examine the case for such a reply and if substantiated would make available resources and/or airtime similar to those offered to the original programme.

The terms under which volunteers work should make it clear that their work, however much responsibility is delegated to them, is subject to a final editorial view by the channel manager.

The kind of programming which volunteers can be expected to undertake is varied. Successful roles for volunteers include all areas of TV production - camera and sound work, directing and presenting, editing, writing, and channel transmission presentation. Few local channels yet have multi-camera studios or OB systems so most programming is likely to be made in a single-camera documentary style. Training here can be quite simple and a relatively high level of indivdual effectiveness quickly achieved. Continuous single camera action coverage or multi-camera action coverage with independent cameras each recording its own output (to be edited later to look like a live OB) requires greater levels of concentration and tighter disciplines.

The range of subjects that local television has already tackled is huge. Perhaps some of the most successful are the human interest stories. Real local people are allowed to express themselves, and those who can perform can be encouraged by public attention. Music and sport are consistently popular with viewers and attractive for a local channel's staff and volunteer teams to work on.

Volunteers have had a go at drama, and in Leicester (for one) there has been an alternative comedy show.

Everyone's regular favourites are children and animals, and the contacts which have to be developed to enable programmes of this kind to appear usually come through the enthusiasm and friendships of volunteers rather than hard legwork on the part of a already harassed channel manager.

This reveals one of the real strengths of volunteer involvement - its ability to break down barriers (real or imagined) between community and company.

Local Colleges

There is one further and potentially important source of local programming - colleges. Many colleges, particularly at the Further Education (FE 16-19 level), have media courses, and resources within the college to produce programming. A word of caution is necessary because such colleges are also likely to be searching for commercial opportunities and their commercial managers will usually be completely unaware of the realities of TV programme making and channel operation.

The relationship between a local channel and local college can develop on two fronts. Understanding the difference is crucial to the relationship. First there are opportunities for media students as producers of real programming - within their courses and in their own time. Secondly there is the development of local educational programming within the local channel schedule.

Neither area is necessarily an easy one for the college to adapt to and any channel manager embarking on this discussion needs to be realistic about the ability of the college to deliver. Media students will not be able to produce 30 minute programmes easily or quickly and to suggest that a college could commit to 30 minutes a week throughout the year is optimistic whatever college enthusiasts may say. They will initially find it difficult to produce five minutes a week. Half-an-hour a month may stretch their capacity without additional staff and technical support, and then there will be no original programming in at least three months of the year.

However, students can become a superb source of volunteer effort and as well as being an ever-willing workforce resourcing all kinds of programmes will be able to make their own programmes largely unsupervised if the local channel makes a commitment of airtime to the college. Well-trained and adapted to the programming concepts which work locally they will be a continuing source for full-time staff as well.

Educational programming requires more expert input. The relationship between teachers and producers has always been a pressure point, and the college needs to recruit a TV professional to manage its own programming if it is to see a real long-term benefit develop. The OCTV model in Blackburn could be adopted for a regular module of educational time on a local channel and this seems a potentially valuable way of a college moving forward with this concept.

The Channel Schedule

Developing a schedule from these small beginnings becomes less of a headache given a structured approach to resourcing the channel. From this debate we have examined how we achieve a news input (even if initially it is in text form only). We have some programming even if only a very few hours a week at the outset. We have a relationship with an educational provider, which will have little immediate impact but major longer-term opportunities, and we have decided to set-up a volunteer group.

To fill our schedule we must ask, is the channel planning to carry acquired programming from within the area or in the national market ?

If so, sources must be found, programmes must be viewed, and copies acquired - either on the appropriate tape format for transmission or on a format which can be handled in the makeup of transmission tapes.

Building this library well before the first transmissions are planned is the recommended practice. Looking at the content and analysing the kinds of programming available is essential. A large city with an active tradition in film and video production (through colleges, the voluntary sector, national/regional arts funding, and/or a local ITV company) may have a huge reservoir of material. A smaller area will be less well-endowed, but resources in a neighbouring city may be applicable to the channel and its viewers - proximity may not be locality but it can be the next best thing.

Making some first assumptions about scheduling based on this programme pool is one effective approach to developing a channel schedule. Once acquired these programmes sit on the shelf waiting to be used. They do not require extra effort to ensure that they are delivered on time (or to budget) and they can offer a benchmark for programme-making standards which staff and volunteers can aspire to emulate.

There is no established ratecard for acquisition of the rights to show such programming. A common approach is to offer £100 for unlimited transmissions over a given period (say two or three years). Such fees are certainly not going to earn a film-maker a fortune but assuming that the programme is already paid for he is being offered clear profit. Copy costs are normally paid by the acquirer.

In the budgetary environment of local television a channel manager will, wherever possible, try to get programming for free. Beware of promotional deals if the programme is the property of a council or a local company - care has to be taken with regard to the ITC's sponsorship rules.

If a film promotes a company it may be entirely fair to charge them for the privilege of showing their programme as an 'infomercial' (ie. a commercial disguised by being in the style of an independent information programme). One way this could be achieved for the overall benefit of the channel is to persuade the company to sponsor a regular original programme in line with the Sponsorship Code. Direct 'quid pro quo' arrangements must however, be carefully handled to stay within the Codes.

Scheduling calls for consistent patterns so that the audience relates to day of week, time of day and programme. Original locally made programming becomes more attractive if it can be placed within or as a series, and its impact improves if it can also sustain a regular repeat pattern. While repeats on mainstream channels are often widely criticised, repeats of local programmes, which are easily missed in the multi-channel ecology of a cable network, are often appreciated. It is not unkown for telephone callers to the local channel to ask for local programmes to be repeated. There is also the opportunity to sell tape copies of programmes.

The repeat pattern can be within the same day/evening, within the week, or across a longer period. A common approach while a channel has limited material available is to 'batch' programming so that two or three schedules cover the week, each schedule being shown several times, possibly on different days. It is also possible within such a pattern to have special schedules for weekends when, if it is technically possible, local sport is an unfailing attraction.

Promotion

The promotion of the schedule is the next problem faced by the channel manager. Often a schedule cannot be finalised in time for promotion through the cable listings magazines sent to subscribers and a separate mailing is costly and impracticable.

I like to draw a distinction between promotion of the channel and promotion of the schedule. Viewer loyalty in the multi-channel environment is not well understood but such evidence suggests that viewers follow channel numbers, time slots and named programme strands more than they do an identified channel purely because it is a channel. Although a local channel will invariably be trying to position itself as a complementary, rather than competitive, element in the cable system's channel line-up the schedule will compete for viewer attention. Therefore getting timing patterns and programme strand names established in the viewer's minds is just as important for the local channel manager as for a national network or satellite programmer.

A succession of one-off programmes under a generic title, looking like a series, is a 'strand' of programming. As long as the schedule demonstrates such 'strands' within a recognisable patterns of slots which are also easily promotable on air the problem of listings can be resolved. The schedule can be promoted well in advance and offers a simple mechanism for informing viewers. Where possible a few lines describing detailed content can be given to listings journals, and if detailed information is not available because of production schedules non-specific information is easily created which identifies a series and motivates viewers to watch a programme.

Sophie's Streets - Sophie Smith visits one of the historic streets of Franchester and talks to the people who live and work there. Could this be the street where you live ?

Club Scene - music and comedy in our pubs and clubs is the heart of local entertainment. Channel 93 joins the Club Scene and brings you spice and sauce from another of Franchester's leading entertainment venues.

Adopting a strand approach for promotional reasons cannot be divorced from the discipline of making the actual programmes.

Wherever possible strand should take an editorial theme, even if it seems very broad. The experience of Canadian local television is that volunteers find working on a strand allows them to develop a personal input within a fixed formula and that offers 'job' satisfaction. Volunteers who seek creative freedom in programme making in return for their involvement, will be less happy if everything they are asked to help with can be seen as formula TV.

Strands should each have their own executive producer charged with delivering the finished product to the desired standard. Such executive producers will have editorial powers and could be volunteers or members of the staff team just as long as the criteria by which they work are clearly established and they are capable of working within these criteria. One key parameter is that they are unlikely to have authority to spend cash, or to make financial commitments. The right to do that will rest firmly with the channel manager.

Once a number of strands are underway with their own impetus building up and their own viewer loyalties developing, a new phenomenon begins to appear - audience feedback. Comments (and appreciation) of the schedule and its programming will offer a regular injection of adrenalin to the channel team. As the channel takes on its own life new programme ideas will flow from this source and new creative skills will come to hand, seemingly from nowhere.

Developing Further

It is easy to get caught up in the on-going life of a TV channel and fail to review what it is doing and achieving. A year into its life there will be good reasons for taking a considered and objective look at what has been achieved, and to be prepared to move it in a new direction. This stimulates some freshness in ideas, can provide a creative spur to the programme-makers (both staff and volunteer) and offers viewers an awareness that the local channel is not resting on any laurels it has earned. It is important that any review also involves the volunteers.

Such a review may also be the opportunity to undertake some audience research and bring this into the review process. Research need not be costly. Simple subscriber questionnaires, on-the street surveys by students from the local college, structured feedback from information providers on any direct responses they may have had, all can be undertaken without commissioning a commercial research company. If a full research project is undertaken it needs to be planned well in advance of the review so that any results can be fed in to the process.

Changes following such a review may be minimal - after all the channel will only be making programmes which it knows it can handle and it knows will be successful with its viewers. But just taking the time to structure a considered review in a formal way of all the channel's activity can offer real value to the operation.

A valid process is to look at the channels Strengths, Weaknesses, Opportunities and Threats (a traditional SWOT analysis). What it has done well, and why. What it has not done, or not done well - and why. What is happening locally, or within the cable company, that might present new challenges, and what is happening generally which can give it problems. A competent volunteer force may need to develop new skills, and understand new disciplines. How is the market changing ? Will, for example, the creation of cable exclusive programming on a new national channel threaten viewing on the local channel at specific times of the day, or might a new national channel become an outlet for the local channel's programming and earn it revenue ?

Whatever the outcomes a successfull local channel will also be asking for more money. The initial equipment may be so hard worked it is now wearing out, new programme schedules may require additional head-end play-out equipment, new graphics computers may offer more visual opportunity which could add value to advertising. New staffing may increase the effectiveness of the channel, and that effectiveness might be visible in extra revenue. The list will go on and on.

This scenario demonstrates an incremental growth which has yet to happen in the UK. The channel which reaches this stage will set new standards for everyone else to emulate.

CHAPTER 9

UNDERSTANDING THE RULES

Licensing local TV and applying the Codes

Television has always been regulated. In the UK growth has been carefully controlled by Parliament with extensive debate at every stage of development. Investigations by committees of the knowledgeable and influential, Royal Commissions and extensive consultancy studies precede formal recommendations through consultative Green Papers and the indicative legislation in White Papers. The introduction of ITV in the mid-1950s was preceded by a six year debate about the advisability of commercial television; the dawning of Channel 4 in the early 80s was preceded by debate about whether it should be BBC3 or ITV2; and cable, with its implied cultural as well as technical threat to an established terrestrial broadcast system has, as early chapters have indicated, always had to accept public debate as a fundamental element in its own growth - such is the nature of the television industry.

The 1984 Cable and Broadcasting Act permitted franchised cable operators to run as many local TV channels in their own licensed area as they wished without taking out additional licences. Content of locally originated channels was regulated and guidance was available from the Cable Authority in very much the same way that guidance for ITV companies was then available from the IBA both regionally and nationally.

The penalty of this generous regime for the cable operator was that he was regarded as the publisher of all the channels that he carried, and by implication, was responsible for cutting out material which crossed any regulatory boundaries regardless of where that TV channel was originating. Bullfighting raised one problem when one network took a Spanish channel. The problem was most evident in material from continental Europe where the UK's 9pm watershed for sex or violence either does not exist or, as their stations are one time zone earlier in the day, their 10pm was the UK's 9pm with consequent UK regulatory problems.

The debate on the 1990 Broadcasting Act introduced a separation between programme licensing and cable network franchising as well as changing the rules for terrestrial television and DBS. A cable network owner can still carry channels from anywhere but if they are not licensed by the ITC or by a regulatory body of a European Union member state the ITC can ban tham. The cable operator can still run its own TV channels, but they now each need their own licence and are regulated entirely separately from the franchise licence, with no cross-over of responsibility. The benefit for the cable franchise operator is that it no longer needs to worry about channel content for channels that are provided by national and European providers.

As long as a channel comes from a European Union country or another state which has ratified the Council of Europe Convention on Transfrontier Television it can carry them without fear of a prosecution for content, and as long as it has agreement for carriage it cannot be regarded as liable for any copyright breaches made by the channel provider.

The division thus created by the 1990 Broadcasting Act goes on to make the responsibility of licensed channel providers quite clear. There is no longer a pre-transmission advisory structure within the regulatory body (now the ITC). The channel provider is now its own regulator. It is supplied with extensive Codes to offer guidelines and if steps are taken over the boundaries it is liable to punitive action by the ITC. If the Commission determines that the boundaries established in its Codes have been breached by a channel, it has significant punitive powers ranging from a formal and public reprimand, to fines of up to £50,000. The ultimate sanction is withdrawal of the licence. The ITC's guidance on the application of sanctions indicates that it would be rare for a fine to be imposed. Warnings have been given for initial offences, sometimes expressed in severe tones, as in the case of Channel 4's 'The Word' in early 1994.

Withdrawal of the licence is something of a strange ultimate sanction as licences for Non-Domestic Satellite Services and for Licensable Programme Services (ie. direct-to-home satellite and local cable channels - not ITV or direct-broadcast satellite) are obtainable relatively easily. There are no franchises and no quality hurdles prior to licensing. There is the implication, though the ITC hopes that this will never need to be addressed, that the management of a licence loser may be declared to be persons unsuitable to hold a further licence, this providing a personal sanction against an offending manager. As with all statutory bodies the ITC's decisions are open to review by the judiciary but it is worth adding that the Commission has, ready to hand, of the best legal advice there is,

The critical issue is to understand the programming and channel operating boundaries and to develop a compliance procedure which avoids problems and which, should problems arise, demonstrates that every best endeavour was used to keep within the terms of the ITC's various Codes. These Codes are sent to every licensee. Proposed amendments are sent to licensees for consultation before they are implemented and complete up-dates are always obtainable from the Commission if required.

Our objective here is to give an insight into the process of obtaining a licence, into some of the thinking behind these Codes and a look at the way they are made to work.

Obtaining a Licence

Local channels require their own licence from the ITC. This is known as a Licensable Programme Service licence (LPS) and is issued, effectively, on demand just as long as the applicant is not deemed by the ITC to be an unsuitable person. There are conditions with regard to ownership of the licensed body (if a company) which limit the interests of bodies with a political or religious objective (including local authorities) and certain other specific potential applicants (eg. the BBC). The wording of the License will be framed to cover a specific service either on named cable networks or nationally.

A small fee is charged for the Licence application and an annual Licence fee is payable. Channels are also required to submit an annual return from which, if substantial amounts of advertising or subscription revenue are being generated, an additional fee based on a small percentage of revenue will be charged.

The Licence carries within it the obligation to observe the various ITC Codes. These are the same codes which apply to Channel 3 and 4 (and 5 if it is awarded). It is the same code that BSkyB, United Artists Programming and the other independent channel operators have to observe. These are:

The Programme Code
The Code on Advertising Standards and Practice
The Rules on Advertising Breaks
The Sponsorship Code

The Programme Code

The ITC Programme Code attempts to encapsulate some of the best traditions of British TV broadcasting and pragmatic interpretations of the intentions of Parliament as expressed in the Broadcasting Act 1990.

The programmer in a cable local channel has to recognise that he has no fall back for decisions outside his own company and if he is the most experienced programme editor available he has a duty to his company and himself to become familiar with all the ITC codes, and the Programme Code in particular. If his channel is a commercial channel and accepts advertising, or seeks programme sponsorship the Advertising and Sponsorship Codes are also compulsory reading.

The whole approach of the 1990 Broadcasting Act is to move programme regulation from a 'before the event advisory and preventative' structure to a 'self-regulating' structure which can administer punitive action by the regulators if the codes are breached.

Some of the terms of the codes are sufficiently vague and open to interpretation so the scope for error can cause concern. However it must be said that in practice the ITC's officers remain of the view that prevention is better than cure and although they cannot be held responsible they are usually prepared to contribute informally to discussions on programme content when a programmer feels that he may be treading close to the boundary of regulated areas.

Issues of balance, good taste and social responsibility lie behind the Code. Good taste is theoretically self explanatory, though whose good taste remains a debating point throughout the TV industry. It is quite possible to interpret the Code such that some of the 'excesses' of BBC TV series, as perceived by the popular press, would be inadmissible under the ITC Programme Code. The BBC is not subject to the same legislation.

It is unlikely that a local channel programmer will find much in the good taste area to concern him. He is likely to be trying to create an appeal for his schedule, so not upsetting the family audience will be high on his list of priorities.

In addition to the ITC the Broadcasting Standards Council has a brief to consider matters of taste and although its powers are more limited the capacity of an intervention to consume large numbers of hours of management time is itself a sufficient disincentive for cable channels to wish to offend.

Balance is the big issue particularly in political areas. Leaving to one side anything to do with elections the underlying rule is that a cable programmer is entitled to request the ITC to release him from the "due balance" requirement of the usual LPS programme licence and replace it with a "no undue prominence" requirement.

This subtle change of wording offers cable channel programmers a safeguard against programme content which may expose a certain opinion or group, to the perceived disadvantage of another. If the "no undue balance" clause is implemented in the channel licence there is no requirement on the programmer to seek a reply, or representation of a contrary point of view. He has no obligation to deliver 'balance', either within a programme or across the totality of his output. He is, however, required to allow a right of reply if requested to provide one, and must do so on terms which are no less favourable than those offered or allowed to the initial programmes. If the clause is not implemented "due balance" requires that the alternative view is sought out and presented on the channel within a timely period, though not necessarily within the same programme as this could present problems.

The coverage of elections is severely constrained by another piece of legislation - The Representation of the People Act 1983. This act supersedes all other legislation when coverage of elections is planned. It has clauses which require equal exposure of all candidates in a constituency, and constrain the timing and style of coverage to achieve an appropriate balance of views on any programme as well as on any channel. In some ways this Act could restrict local programmers to a greater extent than national broadcasters. Local television is much more likely to wish to cover a constituency election and its issues in some depth than it is to discuss national political matters. Considerable care is advised - such issues can take on a life of their own, yet the law dictates that they cannot be separated from the electoral situation.

Coverage of extremist politics becomes difficult within these constraints, especially if a political group tries to obtain coverage for views which can be regarded as inflammatory, or which may incite violence. The channel operator has a duty to social responsibility (and as a matter of public relations will not want to upset responsible subscribers) so the presence of such views can bring all political debate on a local television to a halt.

One of the less well known requirements of the Programme Licence which can affect local channel programmers is that which requires a nominated senior executive to have responsibility for the channel and for compliance with the licence obligations. It appears to be a wish for Managing Directors to have their name on a channel licence but they may wish to reconsider on reading this book. The difficulty is that in the cable industry such senior staff rarely have any editorial or production experience. For a civil engineer or accountant to have the last say in a self-regulating environment, when there may be extremely well qualified junior staff available seems faintly ludicrous. Clearly the implications of breaching the Codes could be severe and it is right that a senior executive should take critical decisions, but there is no training available to them and how they consult with their junior staff is entirely at their discretion. The real danger is that they may be inaccessible at a critical time and if no tight procedural structure is in place a problem could get transmitted by default. It may not do justice to the programme makers who have de facto day-to-day control and interpret the Codes as part of their daily role. When significant fines are the penalty for breaches of the Code it is understandable if directors feel that retaining nominal control is playing safe.

The ideal would be for a senior member of the company management team to have a line responsibility for the local channel and to have sufficient time allowed for an involvement in decision making if necessary - while allowing the channel manager to be the one who actually takes the routine decisions. Formal procedures would nominate deputies and make them aware of the times when they were responsible. All those involved could be named on the licence and be empowered through these formal procedures to take responsibility.

This way those working on the channel, whether staff or volunteers, can have confidence that someone with appropriate clout is always available if necessary, but can also feel that proper delegation will make that situation unnecessary. This is important for morale. The executive without programming experience and who needs to get involved must be ready to learn, listen and discuss, and not be over ready to deliver glib judgements. His perceptions of the code may not accord with the experience of programme or advertising issues that his team can bring to the discussion.

There is also the need for a formal complaints procedure to be in place and observed. This requires that calls, letters and replies are properly recorded so that should an investigation be necessary by the ITC or the Broadcasting Complaints Commission their task can be undertaken with minimum disruption. The BCC will only get involved if a programme is seen by a complainant to have unjustly or unfairly treated him or to have invaded his privacy. This applies even if the programme is made with his knowledge. It is worth remembering that the civil law with regard to libel applies in full to local television programmes.

Local programme makers can feel that they should be prepared to be outspoken, and to challenge their local community, if they are to attract viewers. Blandness is not a great audience grabber. In this light the regulatory and legal structure can be seen as unpromising. But responsibly handled outspoken-ness can be encouraged. The force of the Code and the law will only be used when good practice is deliberately flouted or irresponsible behaviour is allowed to reach the viewer.

The need for a complaints procedure is a sensible procedural safeguard as well as a licence requirement. A sample procedure is printed overleaf and can be adapted for local terminology as necessary. If a procedure is not in place, or not implemented despite being publicised, the channel operator could be held more culpable by the ITC, the BSC or BCC if a viewer makes a formal complaint to the regulators.

It is a fact of life that more programme complaints are settled by a simple letter or phone call from front-line management than ever reach the regulatory bodies. If complaints reach them unnecessarily, because procedures are not in place, they will have every right to be aggrieved.

Advertising

The key underlying principle of the advertising code is that all advertisements should be truthful and accurate in their representation. Claims which cannot be justified under examination are not allowed. Complaints against advertisements occur daily and the ITC has a regular publication to licensees advising them of the latest decisions.

Examination of claims can be extremely detailed. Academic and industrial experts are occasionally called upon to examine highly technical claims made in advertisements and challenged by viewers. The legalese of detailed judgements can be arcane, but this necessary process establishes detailed precedents which licensees need to be able to follow.

A PROGRAMME COMPLAINTS PROCEDURE

1. If a complaint about the local channel is received it will be logged with the name, address and telephone number of the complainant and full details on the Complaint Form. The complainant will be assured that a manager will contact him.
2. The Complaint Form will be copied to the Managing Director, Marketing Director, Channel Manager and the Programme Complaints file.
3. The Marketing Director will appoint an investigating manager who will contact and establish the full details of the complaint. The complainant will be advised of his/her right to contact the ITC and/or BCC and will be given addresses and phone numbers. A written record of the conversation will be filed in the Programme Complaints file and a copy sent to the complainant.
4. The investigating manager will investigate the complaint as appropriate and consider whether there is a case for action to be taken. A written report of the investigation together with any recommendations will be made by the investigating manager. A copy of this report will go to Marketing Director, Managing Director and Channel Manager. A copy, together with any supporting documentation, will be filed in the Programme Complaints file.
5. A copy of the report will be given, in person by the investigating manager, to any individuals (staff, freelance or volunteer) against whom disciplinary action may be considered as a result of the complaint being upheld.
6. The Marketing Director will confirm what action is to be taken as a result of the investigation and advise the complainant accordingly. The complainant will also be advised that he can ask the company to review the case, or can make a formal complaint to the ITC or the BCC. A written record of the contact with any supporting documentation will be filed in the Programme Complaints file. In normal circumstances this stage in the investigation will be reached within 5 working days of the complaint being received.
7. In the case of a requested review the Managing Director will review the file on the complaint and determine a final ruling from the company. The complainant will be advised of the decision and a full written record placed in the Programme Complaints file. The complainant will again be advised of his right to take the complaint to the regulatory bodies.
8. Any disciplinary action recommended following the investigation against staff, freelancers or volunteers will be taken in accordance with the appropriate disciplinary procedures.

The Broadcasting Advertising Clearance Centre is run by the Independent Television Association (ITVA) on behalf of the Channel 3 and Channel 4 companies and BSkyB. The Centre was originally the copy clearance department of the ITVA is the main source of expertise and experience on standards in television advertising and cable channels with a regular TV advertising throughput are well advised to seek a clearance contract with them. This is not likely to be necessary for simple graphics adverts (all that many channels will aspire to).

Even for the simplest adverts the critical eye must remain in action. Claims using superlatives (best, cheapest, most ...) should always be questioned. Product comparisons with competitor products should be treated with extreme caution. It should be usual if any kind of claim is made in an advert for the advertiser to justify his claim in writing. Examination of this justification can then be made and advice sought. Reference to indepndent confirmatory opinion may be made or, if the sums involved are small, it may just be easier to reject the advert.

Sponsorship

The 1990 Broadcasting Act allowed programmes to be sponsored. There is a great attraction for a major national brand in sponsoring a high profile programme on network TV and large sums can be raised towards production costs.

In the local market there are fewer examples. We have already mentioned the Coventry pub quiz series. It was sponsored by the Midlands brewery M&B. The aim was to use the quiz transmissions in its own pubs and an entry form was also inserted as an ad in the local evening paper on the day of transmission, thus allowing every viewer to enter the competition. The brewery received logos at the top and tail of the programme and the mid-programme advertising breaks in line with the Advertising Code. It was also able to promote its involvement with local TV through ads associated with the entry form printed in the newspaper. The Code does not allow them to place product, or logos in the programme, nor can the programme mention the brewery or its products by name. However the pubs taking part, as independent businesses received their own on-screen promotion for their part in the quiz programme, as an advertisement paid for by the brewery, at other times earlier in the day.

The whole exercise became a revolving piece of local economic activity benefiting pubs, brewery, programme makers, TV channel, newspaper, the local club comedians who hosted the programme, and the weekly winners of the quiz who received cash prizes.

Sponsorship needs to be carefully managed, using the Code for guidance at all stages, and referring to the ITC where in doubt. The Advertising and Sponsorship Division is always glad to help.

Religious Programmes

Christian television was singled out for particular debate in the passage of the 1990 Broadcasting Act through Parliament. There were real fears that a multi-channel explosion in the 1990s would allow the evangelical miracle-working fund-raising moral majority explosion of the mid-80s USA to be exported here.

The result is that religious programmes are not allowed to fund-raise or to make claims about their religion, although personal statements are permitted. Comparisons with or comments about other faiths are not allowed either.

It all seems faintly ridiculous now that some of the US 1980s Christian 'stars' have been shown to be shams but it is a genre of religious programming which commits and attracts large audiences across the Atlantic. The prospect of a fringe cult managing to slip through the regulatory structure by default, and breach the regulations despite being made aware of the Codes, is not an idle threat. There is evidence already that attempts have been made. Local channels may find it easy to ignore religious programming in fear of stepping outside the Codes but this would be a short-sighted approach.

A local consultative group may offer a way of ensuring that any proposed religious output comes within the guidelines. The make-up of this group should be a proper cross-representation of the religious life of the franchise area, though it will inevitably be dominated by the mainstream churches. The Churches Advisory Council on Local Broadcasting, set up in the 1970s to help resource local radio, now has a local television consultative group and can be approached for guidance.

The potential for Christian groups to provide regular programming is high. Early work in Coventry, Blackburn and Leicester has shown that the churches working together can become an important resource for local channels. Because these churches come from the mainstream they are unlikely to step outside the accepted and published codes of behaviour. The attraction of local TV for them is as much about serving the need of the community to discuss major issues as it is to proselytise through programmes of worship or preaching.

Summary

The Licence structure and the ITC Codes are the main evidence of Parliament's intention to have a 'light touch' regulatory structure for TV in the 1990s. Leaving responsibility for front line regulation to channel operators does open it up to abuse more by default than intent.

In return for the 'light-touch' the ITC has been granted some fairly severe powers, which it intends to use with great discretion. However it is worth any local channel manager remembering that the ITC does have real powers and in an extreme case it will use them.

Let us hope that time never comes.

CHAPTER 10

VOLUNTEERS IN CABLE PROGRAMMEING

Cost saving, creativity and caution

During 1993 the shape the UK cable industry developed was one based on 'super-franchises' and there is every reason to suppose that, as the next round of franchising gets underway from 1994 onwards, the size of franchise blocks will increase.

The super-franchises are ready-made for a form of regional television and the award by the London Interconnect group of cable companies in late 1993 of a contract to Channel One Ltd, owned by Associated Newspaper and SelecTV for a London channel is an indicator of how large franchise blocks are likely to think.

The effect on local television of regional interconnect channels is hard to calculate, and whether the ITC will willingly forego any pressure it might otherwise put on when an interconnect channel is started is also unknown. One effect seems certain. Local channels in such areas will face more competition and the need to develop uniqueness will be more evident than ever before.

Whether facing competition from an interconnect channel or not, whether run by a cable company or an independent, whether maintaining a heavy transmission schedule, or simply inserting material for a very few hours a day, local channels will eventually be heavily dependent on volunteers.

Regional interconnect channels will be organised more along the lines of a regional broadcaster, local channels will necessarily recognise that their credibility and viability relies on locally produced programming. Staffing costs are the biggest single cost in a TV channel operation and a significant volume output is only possible if the main output of local programmes is made or provided on a nil or extremely low-cost basis.

As any channel will want to have as much local programming as possible so that it can demonstrate the value it adds to the community (and to the cable operation) it is inevitable that volunteers will play a growing role in local television.

Obligations and Responsibilities

Using volunteers entails obligations, as well as bringing advantages. One issue that will quickly come to the fore is 'exploitation' - especially if the channel operates on a commercial basis attracting advertising and sponsorship. Another critical factor is "who owns the programmes " ?

There are also many practical issues - insurance, liability, security, editorial freedoms, financial accountability etc. There may be issues of unemployment law and associated National Health and Social Services implications and, as we have seen from the case studies in Chapter 7, there can be personal agendas as well.

Starting a volunteer group is not something to be entered into lightly. It has to be undertaken as a managed and documented exercise with written consents from volunteers and a clear understanding of what they can and cannot do or say when they are engaged in programme making. They have to be trained and the training has to be virtually continuous.

In time volunteers can also become trainers. The trusted few who reach this stage will also be in line for recruitment when the staff team is ready to expand, or needs to replace staff who leave. More than one channel manager in Canada is a former volunteer and there has been at least one example in the UK already.

There are real employment opportunities for volunteers who take the role seriously, a fact which itself can be promoted.

Adding Value

The whole purpose of starting a volunteer group is to add value to what is already achievable. Their value is expressed as local relevance and quantity - hours of programming that local people want to watch. The concept that must be accepted that the value that volunteers bring to an operation is not a cost-cutting approach, nor is it simply a method of implementing community access. It is a way of creating an audience perception of value in the channel and improving the all round variety and content quality of programming without being exploitative, and thus satisfying the (well researched) wants that a community has of local television.

If a profit results directly from volunteer involvement it can be returned to the community in an appropriate 'charitable' way, or re-invested in supporting volunteer activities.

Other benefits of a volunteer group include a freshness of ideas and a readiness to challenge conventions, knowingly or unknowingly.

Creativity

The new talent and challenging creative energy in broadcasting in the last 25 years has tended to come from two quarters - the young and the local. Pirate radio in the 1960s and then local radio in the 1970s broke the conventions of radio and created new styles of broadcasting. Channel 4 challenged the conventions of television organisation and although it took some time to severe the stranglehold of the television unions, the young gained access to the national TV screen in a way that had never been achieved before.

The radio pirates waged a creative and imaginative war on the BBC during the 1960s which threatened the soul of national music radio. The much vaunted launch of Radio 1 in 1968 managed to stem the flow - and it attracted into its fold and 'legitimised' some of pirate radio's most experienced and talented disc jockeys and producers. The creation of independent local commercial radio saw the creative constraints break down and the chains disappear altogether.

The local lads and lasses who communicate to radio audiences, sometimes as small as 4-5000 in number, in towns and cities across the country are the heartland of local radio communications. They work in structures which are less cumbersome than the local press, and because they work in a 'live' environment the editorial constraints are the ones they themselves create, not ones which are imposed by older, and supposedly wiser heads.

It is this environment that local television will eventually tap into. The energy and enthusiasm which has driven radio from the county or regional base, to a town or country district base, is the same energy and enthusiasm which local TV must tap.

We have seen the signs of this energy already. The precocious nature of Cable London's programming in 1992 came from the professional end of this spectrum. Many of the CLTV volunteers were in fact TV professionals, made redundant by Thames Television or TV-am and seeking an outlet for their enthusiasm and skills.

We see it too in Leicester's Cable 7 channel, where people who have never handled a video camera before have quickly challenged TV conventions and have asked, for example, why a sports programme is so often handled the way it is.

This questioning is the same that has driven much of Channel 4's output. 'The Big Breakfast' is hardly conventional, its presenters are young and do things such as addressing the camera, calling off camera (and out of its range) to members of the audience or the technical crew, and generally conduct the programme in ways which those brought up to the strict conventions fostered by 'In Town Tonight' and 'Nationwide' would find disturbing.

One of the team, Mark Lamarr, made his TV debut on Cable London's local channel, as presenter of one of its first regular programmes - 'Diary'. He gave a singular viewpoint on an area within the franchise, calling on local people in a casual way to reflect on what made their area different. Its very first programme was shot within 100 yards of Kentish Town tube station - hardly a notable location in national geography but at that time at the heart of the area served by the cable network. Reflections on how the quality of vegetables have changed over 50 years is hardly the stuff of network TV, but handled in his style with local relevance it became watchable TV.

Clyde Cablevision managed to create an outstanding record locally for fresh programming approaches and several of its former staff, volunteers and on-screen artists have moved into national television.

Local television is a good breeding ground for new creative ideas in programming. It offers opportunities to volunteer programme makers, and an opportunity for the talented to try out their ideas and 'stretch their wings'. When they fly the nest no-one should be surprised and the channel should not attempt to re-create their successes, but should be able and ready to move onto new things. Volunteers can be the lifeblood of creativity.

Management of Volunteers

This is the talent that local TV must foster but it must do it in its own way, not as a reflection of radio.

Has local television the access to management skills which can achieve this breakthrough ? Are there any training courses for managers of volunteer groups ? Do senior cable managers actually recognise the value and worth of the skills they build in their local channel managers, and in the volunteers ?

It is significant that even when confidence is shown in one

person to exercise high programme judgement skills at a junior level, the eventual senior manager may be prepared to exercise low programme judgement skills at a senior level, and outweigh any innovation, or good, or confidence created in voluntary contributors. Programme judgements on creative issues rely as much on an instinct born of awareness of programmes as craft items, as from any conscious editorial judgement. Such judgments are difficult to make if the person who makes them is an engineer, or a salesman, and not trained to make them. Programme creativity is fostered by a combination of trust, openness and mutual commitment and cannot be reduced to a chart, or measured on an oscilloscope. The BBC and ITV companies learned that lesson years ago.

Cable companies with local channels and volunteer groups must ensure that their managers are properly trained and then must trust them. Decisions about suitability of material for transmission must be made according to known and agreed criteria (within the ITC Programme Code) and managers must be trusted to work within these criteria. Implications for the output if this is not allowed to happen could be significant. Volunteers do not have to comply with something if they do not want to, and if frustrated can walk away from the channel and from programme making, even of programmes that they themselves created. At the same time spoiling them like prima donnas will not achieve anything; requests will become demands and the demands intolerable.

In the days of a very left-wing Camden we had some focussed discussion with elected representatives of the voluntary sector on how local programming could be managed. In principle there was little difference between our beliefs, but in practice there was a substantial distance to cross. The voluntary media sector is used to a high degree of freedom, and it must be said generally exercises a high degree of responsibility in the delivery of local services (for the disabled, unemployed, housing etc). They could not readily come to terms with the idea that whatever they contributed in voluntary effort to cable programming had to be subject to editorial control from the company (there was no alternative under the 1984 Cable and Broadcasting Act). Even the creation of a jointly managed 'trust' with a cable company employee as editor-in-chief was not an idea that could be easily accepted.

The idea of such a trust also raises the possibility of the channel insulating itself from direct contact with volunteers by working through contributing producers who take volunteers under their own management. No channel manager will be unhappy with the idea of volunteers working on a strand of programming from a local programme provider and being managed directly by that programme provider provided they are assured that the volunteers behave responsibly and do not create poor impressions of the channel itself.

How these contributing producers are themselves managed (or paid) is another question. The model for this already exists in Coventry. Revd Lawrence Mortimer, the communications officer of Coventry Diocese, has a volunteer group who make a weekly Christian programme for the local COLT Channel and who rarely individually come into contact with the cable company. In this instance Lawrence is a staff member of the Church of England and one of his duties is the TV programme and the training and management of the resources it uses.

The Issues

One of the first things volunteers have to understand is that they are not employed by the local channel (if it is a company in its own right) or by the cable operator. The distinction between "working for" and "employed by" is a subtle one even in law but it is critical that the volunteer accepts and understands the implications of the fact of being a volunteer.

These implications should be laid out in a simple agreement which the volunteer signs to accept. The 'reward' for acceptance is that that of becoming a member of a formally constituted group - whatever it is called. There may be a membership card and a membership fee.

Training

His first obligation is to get trained. The channel operator may offer this training either in-house or working with another agency (such as an FE college). On completion of training, and possibly while undertaking training, the opportunity will arise to work on programmes. The Canadian model leaves it to the volunteer to make it clear to the channel management that he/she is available for such work and gives them the information from which they can make their own decisions. A trained volunteer who sits at home and waits for a phone call is likely to wait a long time. This works well in any volunteer arrangement, the well-motivated are the ones that the channel actually wants to work with.

Health and Safety

There are few rights attached to being a volunteer but one that does exist is basic health and safety legislation. A volunteer cannot be placed in a more hazardous location than a staff member, but even more importantly a volunteer cannot be allowed to voluntarily place himself in such situations. The right also carries responsibilities - volunteers have a duty to behave in a manner which demonstrates proper regard for their own safety as well as that of others. The camera shot may be brilliant but the risks are the same, staff or volunteer, and the penalties for the company could be the same if there were an accident.

Insurance

Insurance is a natural follow up subject. Volunteers are not employees and will not therefore be covered by the normal corporate insurances. The company's insurance broker should be consulted and the cover which is available should be advised to volunteers. Normally a personal insurance would cover them for leisure activities but the sad fact of life is that many volunteers are unemployed and cannot afford such insurances.

Accident liability is probably the biggest problem. If a volunteer drops one of your cameras onto the head of a passer-by while he is out shooting one of your programmes, who gets sued ? The answer is simple, you do; because lawyers will assume that the individual is unlikely to have the resources to make it worthwhile taking action against him !

If the editorial control processes are properly implemented and the frontline editors properly trained there should be no legal liability problems from programme content - through libel, defamation etc. It is worth stating that good procedures and properly trained staff will be easier to insure than a channel which acquires a record of indiscretion.

Insurers will also be interested in your equipment. It is quite likely that they may demand that your kit only goes out with a responsible and trained member of staff and require that person takes full responsibility for the equipment in use. This is a quite understandable restriction, from an insurer's point of view, but may be unworkable from the viewpoint of the manager of a cable channel. There may be a good case for a set of low-cost sub-professional location equipment which goes out on an uninsured trust basis (possibly with a small deposit to encourage a responsible approach from the volunteer who draws it from the stores).

In all these areas risk, reward and cost have to be balanced. The final equation will depend on the advice available and the stance a management wishes to take. Whatever route is taken do make it a conscious set of decisions, and do not let these issues go by default.

Representation

No volunteer can be allowed to 'represent' the channel or the company except as specifically authorised. The authorisation (for example to visit a venue and plan coverage of a sporting event) should be specifically laid out in a letter which can be presented to the location or event organiser, and on which they can refer back to the channel manager.

The representation issue can be important should a programme become contentious. The local press may well search out people who worked on a production to seek confirmation of a story, or gather an opinion. It is easy to imagine the headlines which could result from an unguarded and innocent comment from a volunteer who worked on a programme and was unaware of any contention which surrounded it. As a fall back position the channel needs to be able to dis-associate itself from such comment.

Copyright

Copyright is being co-ordinated into an international code backed by law in every major country. A contributor who makes a creative input into a programme could (although generally it would be unlikely) make a claim to some intellectual or moral rights in it. Although moral rights are not recognised in English law, other European countries do recognise them and European law is being 'unified'. It is important that volunteers formally relinquish all rights in any programme on which they work, by assigning such rights in a blanket way to the channel.

Where productions are wholly volunteer produced, even to the extent of covering their own cash expenses, and the only input from the channel has been the loan of equipment, there is a good case for a shared ownership approach.

This would allow the cable channel to use the programme in an unrestricted way for its own transmissions, and maybe as a swap with other local channels, but would allow any sales to generate a royalty payment to the volunteers involved. This can also encourage the volunteers to create a quality product which may have a secondary market.

Copyright also applies to local people who appear in programmes. A form which acts as a 'release' from all rights in their performance should be signed by every contributor. This even includes professionals (musicians, actors, personalities) who may be appearing for nothing. They are, however, likely to want to restrict their 'release' to local output only and not accept secondary usage of the programme in which they appear without some form of payment.

Exploitation

The issue of exploitation is an immediate and natural follow-up issue to such an agreement. How far can the channel go in selling programmes, exchanging them, or adding revenue raising activity (such as advertising) around them when the major cost element, labour, has been volunteered ?

Of all the issues which volunteer groups raise this is possibly the most complex because it raises questions of morality which have so far been unexplored in the media. One-off instances of volunteer commitment to programme making in broadcasting are far removed from a contuing commitment around which a local channel would be built.

Canadian local television has solved the problem in a simple and pragmatic way. Community channels which carry programming, whether made with volunteer assistance or not, are not allowed to sell advertising or sponsorship. Commercial local channels can do so, and almost every Canadian cable network has at least one channel with wall-to-wall adverts, 'infomercials' and sponsored slots. Quite often the commercial channel raises sufficient cash to leave a profit after the costs of the community channel are deducted, although in the accounts (and the minds) of the Canadian cable operator this is not a critical equation. In Canada the community channel is funded through a precept on the basic cable subscription and is enshrined in Federal law. Therefore in practice the profits of the commercial channel are straight profit.

In the UK each cable operator is free to approach this issue in the way he desires. It is possible that we shall see Canadian owned companies adopting the same approach as in Canada - thus avoiding the moral dilemma.

Other channels, particularly 'third party' channels which may use volunteer effort, have to be able to answer any challenges their volunteers may make on the question of exploitation. It is unlikely that anyone will wish to create the situation where a volunteer group will publicly withdraw its support so openness and trust apply as much here as anywhere else.

There is no 'magic formula', nor is there a precedent which can be reliably quoted, as no-one has yet had been forced to face this issue in practice.

An argument which does carry weight is the concept that volunteers are investing in their local community, rather than the TV channel, by committing their time to programme making. The investment is both social and economic - helping to generate a better community through information and entertainment, and through increased economic activity such as advertising. There are those for whom this formula will not carry any weight.

An appropriate practical route to segment the output so that distinctions between programme sources are clear may be to create slots on the channel which are free of advertisements and in which only volunteer produced community programming is transmitted. It may reduce the overall commercial potential of the channel, but (arguably) the programming watchability is increased, thereby gaining advantage for the commercially supported programming which is carried. Inevitably some peak hour time is likely to be allocated to such programming. As this is the most attractive period for advertisers the sales team on the channel may be less than enthusiastic. A compromise may well be to actively seek local sponsorship for peak hour slots.

Another approach may be active commercial consultation with volunteers - advising them that in order to create a production, sponsorship or advertising must be acquired. Some kind of 'reward' for their support (and their 'selling' effort) may be offered, but the income tax or the employment implications could be significant - particularly for anyone in receipt of state benefits. It is possible that the financial reward may be for the benefit of the volunteer group as a whole, or even become community focussed benefit such as support for a local good cause or charity.

Whatever the route chosen to bring in revenue alongside a volunteer produced programme the issue should be approached in a considered way well in advance. Consensus agreement to any proposal may not be easy to gain and time may be required to find sufficient volunteers prepared to accept a chosen course of action applied to a particular programme.

In some cases the best mutually acceptable route may be production on a commercial basis, with volunteers receiving payment for their contributions. Do take advice from the company's employment experts before committing to this route. As indicated above some of the volunteers may open themselves up to loss of state benefits, or additional tax demands.

It is easy to ignore this issue, and believe it may never happen. The dangers of doing so are obvious. It is not an easy question to grasp, but is an important one if volunteer effort is to be sought.

Summary

There is a lot of excitement to be gained from working with volunteer programme makers. The rewards for the channel can be great. The issues, however, must be addressed at the outset because a failure to do so could leave a channel stranded, and its owners liable for a nightmare of litigation from all directions. Naivete is no defence in law, and no excuse if a channel falls off air because its production support has walked away.

CHAPTER 11

TECHNOLOGY FOR ALL

A pathway for technical development

Cable television in the UK has no specific technical regulations which its channels are required to observe. The standards which a cable network regards as acceptable are for its own management to determine. The concept of 'broadcast standard' as administered by the old IBA and ITA before it, only exists for the terrestrial broadcaster. Satellite broadcasters work to the highest practicable standards they can achieve for the type of programming they are making.

By contrast it is fair to say that most local television channels on cable systems in the USA and Canada tend to work to the lowest technical standards they can get away with. The likelihood is that a similar route may be adopted in the UK but one reality of the 1990s is that the cost of video production and post-production (editing) technology is rapidly dropping and new methods of achieving programmes of a high technical quality are appearing with capabilities and costs to suit every ambition and every pocket.

In this chapter we look at some of the issues which lie behind the choice, considering different technical standards and why choices may not be as straightforward as they might, at first, appear. We shall also look at the quality of support which is required to make any technology work satisfactorily in creating local programmes.

Photo-Videotext and Teletext

We have seen that a cable local channel is likely to get underway with a relatively simple text/graphics system. This will usually comprise a graphics computer (with software), and a head-end server possibly with the addition of a teletext inserter.

Teletext is the text only system transmitted by most TV channels in a number of lines in the 'vertical blanking interval'. A typical insertor will have the capacity for 250 or 500 frames of text data which is computer generated and downloaded into the insertor. A simple interactive structure can be devised using the page numbers which are available. The resulting text can be seen by any subscriber with a teletext equipped TV set.

Photo-videotext occupies the full capacity of a TV channel, but can have teletext running 'behind' it just like any other channel and the content of the two can be linked using the teletext page numbers. Deciding on the graphics computer is a question of assessing what the needs are and specifying hardware and software to meet those needs. Suitable systems today are based on personal computers and several makes offer alternatives. But beware of easy, or cheap answers. The needs definition should be thorough and the seductions of technical niceties demonstrated by eager salesmen ignored. Failing to take experienced advice, and thus forgetting (or not discovering) a key element until it is too late, helps no-one - least of all the channel.

The needs which are obvious are about the graphic images, less obvious is that the library of these images will grow rapidly and will need extensive management. A key issue in this management is scheduling. No-one pretends that a subscriber will sit down and view hour after hour of graphics. The objective of the channel is to create a presence and offer information which viewers can access at their discretion. Because full screen information cannot be supplied on-demand the scheduling of specific segments of information at specific times helps the channel operator's promotion of the channel to the audience. Without it he only has the dumbest of channels and creating an identity and sense of value for the audience is much more difficult.

Television Production Equipment

Full broadcast standard video production can be reached with Betacam SP video, or all-digital videotape formats of which there are now a wide variety.

Until very recently all these routes were costly. Even with prices dropping as rapidly as they have been in recent years buying the equipment to run a TV channel requires budgets of tens of thousands of pounds. At the higher broadcast formats there is also the need for a level of full-time professional engineering support if equipment is to be maintained to a consistent standard which delivers the quality of which it is capable.

Sony's Pro-Beta and JVC's MII formats can be found at a lower cost and Sony also supplies a lower-cost version of Pro-Beta (UVW) aimed at the market which has been dominated by the older U-matic format for the last 15 years (including cable). Both Pro-Beta and MII are widely accepted in broadcasting (if unofficially) and deliver pictures which, to the uninitiated, are impossible to tell apart from those created by the full broadcast formats. Editing in these two formats will deliver programming largely indistinguishable from broadcast Betacam SP (some argue that MII is better) but at a far lower capital outlay even allowing for a similar degree of equipment automation.

Sony's Hi8, and SVHS from JVC, Hitachi and Panasonic are the next formats on the scale down from Pro-Beta. Both offer excellent picture quality and a choice of equipment from consumer product up to flexible professional standard systems. Mixing these formats with the Betacam UVW format may prove to be an effective high-quality low-cost route forward.

The older U-matic format (including Low-band U-matic, Hi-band U-matic, BVU and BVU-SP, and U-matic SP) is now reaching the end of its life and will only generally be considered if it is necessary to buy second-hand equipment. U-matic format equipment will be around for many years to come. For most of the last 20 years it has been a mainstay of non-broadcast professional production, reflected in the huge numbers of corporate videos, educational TV series, and local cable TV programmes that have been made using it all around the world.

Image Acquisition

It is in image acquisition that the lower formats such as Hi8 and SVHS gain their greatest advantages for local television production. Portability and quality at low cost allow production tools to be put into more hands, and lower tape costs have all but eliminated the broadcasters age-old worries about the per minute cost of the stock to record original pictures (shooting ratio).

High-band 8mm (Hi8) and Super VHS (SVHS) are widely used by broadcasters, even if they make little announcement about it, particularly in news-gathering, 'fly-on-the-wall' documentaries and special or risky situations. The biggest disadvantage of these formats is the way that sound is handled. The main audio signal in Hi8 is recorded in an interlaced format with the picture, and cannot be re-recorded without being taken out to a mixer during editing. This may not be a problem in many simple productions but when multiple sound sources have to be mixed during editing it can add complications. The alternative is to record audio in the spare space at the side of the tape (PCM audio). This is only possible in camcorders equipped to record it, and inevitably these tend to be higher cost, professional equipment.

Domestic level cameras usually have the ability to record sound from either an on-camera microphone or a plug-in mic using a simple mini-jack connector. Cheap add-on microphones can give an unexpected good quality but there is the danger of radio-frequency interference in the cable between mic and camera. There is also the question of how good a microphone should be used, even when sound can be a critical part of the production. The quality of the sound being recorded will be affected by the quality of internal audio circuitry, connectors and the recording heads on the camcorder being used. It could be a complete waste of effort attaching an expensive high quality microphone to a £1000 camcorder.

On consumer camcorders audio level control will be automatic. This creates a 'wow' effect when natural background noise is high - sound close to the microphone causes the background level to drop, but when talking ceases the automatic level circuit brings the background noise up to a foreground level. It is virtually impossible to correct this in editing and it can be quite disturbing for the listener.

Two big disadvantages of domestic type cameras are the setting of white-balance and the provision of automatic focus. White balance provides for automatic processing of the colour in the image being recorded so that whites are always white (everything else follows). If conscious decisions are not taken each time the shooting environment is changed it is easy to get scenes with a green cast (fluorescent light), orange cast (tungsten light) or blue cast (daylight). It is possible to correct the colour balance with specialist equipment during post-production but the essential white reference is not there and adjustment becomes a matter for subjective decision-making.

Automatic focus may be wonderful for off-the-cuff shots at a wedding or kid's party but it is critical that it is not left on during shooting something planned and considered. It can be a useful tool for setting focus on an interview (for example) but should then be turned off. The effect automatic focus offers when it 'hunts' for a correct setting can also be quite disturbing, and immediately gives the production an amateur feel.

Another problem which can occur is the inability to adjust shutter speeds to computer screens which are in shot.

Finally high contrast and low-light images will have less detail and chromatic range when shot on consumer cameras with single-chip image systems, than will happen on the three-chip professional systems.

Using low-cost consumer cameras should, therefore, be carefully considered and no channel which aspires to volume output will be likely to regard them as the mainstay of its production equipment. Nonetheless these camcorders have a real value for loan to volunteers although the potential for success in the likely environment in which material will be shot should be assessed before each production using them. Returning with beautifully composed images but with sound which is difficult to edit and poor colour balance, will make a production as unusable as anything else which seems more obvious at the time of shooting.

Professional Camera Systems

Professional Hi8 and SVHS equipment offer more comprehensive facilities for handling sound. Sockets will be of a professional standard (usually XLR or standard jack) immediately implying the use of better quality microphones, the ability to plug in a portable sound mixer taking control of sound levels away from the camera, and the ability to monitor sound on headphones as it is recorded are both highly desirable features. Some cameras also allow recording on different tracks, and have on-camera indicators that all is proceeding correctly.

The top of the range Hi8 and SVHS equipment is professional standard and like the higher formats requires specific training before it can be used effectively. There is a growing emphasis on automated controls with manual over-rides on professional camera systems and these can add speed of operation and convenience to the cameraman, especially when he may also be covering other roles in a local TV production.

The two professional Betacam formats are only marginally more expensive than the top of the Hi8/SVHS ranges. The UVW range of equipment is replacing U-matic (although this 20 year old format is still doing good service) and it offers a more consistent quality, a better resolution in average shooting conditions and at prices which are only a fraction more costly than Hi8. Its overall performance in post-production is far superior and this, if nothing else, would give it a natural edge for a programme making channel.

Mixed systems are quite usual. It would not be strange to have a mix of Hi8 and Pro-Beta in a station's equipment. This would require a conscious decision to be made about which formats to use for which programmes, but would allow any budget to stretch further by using Hi8 for volunteers, simple event coverage and adventurous situations, while Beta was used for music, complex productions and as the edit master format for all editing. The compatibility between machines from a single manufacturer is excellent and edit equipment can be changed over from Hi8 to Pro-Beta without any major problems occurring.

Post Production

One of the worst things that has happened to the development of television is the invention of cheap visual effects devices for use in post-production. The importance of the 'cut' - the simple clean changeover between two shots - is too easily forgotten when the attractions of wipes, mixes, and flip-flop effects are presented to the amateur editor.

It is worth looking at top quality broadcast programming to remind yourself of the value of the cut. It is possible to see whole hours of TV where the use of the cut is skillfully exploited and other transition effects are almost non-existent. The lesson for a cable system establishing itself with (possibly) a high proportion of relatively unskilled staff, is that two-machine video-editing (where the only transition that can be made is a cut) should be the backbone of post-production. In Canada there are community channels now decades old where all editing is still undertaken on two-machine systems, and the use of mixes or dissolves is confined to studio coverage, whether recorded or live.

Two machine editing systems are relatively cheap to establish. My preference would always be for 2 such editing system over a single 3-machine system - the cost is about the same. The additional throughput from 2 edit systems more than compensates for the loss of effects and the supposed productivity that a 3-machine system offers is often lost by the indecision of the producers directing the edit.

The three machine (or more) edit suite does have other values apart from the ability to create additional video effects. The main one is that when a production has multiple locations, recorded on a variety of different tapes, they can be loaded and set-up independently of the editing, and while editing is going on.

Non-linear editing is now a growth area in post-production. Using a computer with digitised video sequences, decision making can be undertaken on the desktop - even at home. An edit list is produced (using time code references) and this list is then matched with the original production tapes. In the more advanced (and currently more costly) non-linear editing systems (eg. AVID and Heavyworks) it is possible to digitise the images to transmission standards in the editing computer. The editing process (including multi-track audio post-production and the addition of complex image transition effects) can be undertaken in the one piece of equipment - finally outputting the finished result to videotape for the actual transmission.

The introduction of digital play-out equipment which will take these digital files straight to the network is already taking place, though the costs are likely to limit their use for a year or two yet to the major broadcasters.

Graphics

A key component of any television station is the graphics. There will be a need for graphics within programmes but there is an even greater need to support the 'interstitials' - the gaps between the programmes.

Cross promotion of other channels, trailers for later programming, a promotion for the cable operator's telephony services, a station ident, lead-in and lead-out idents for advertising breaks, 'what next' information, sponsorship logos, the local weather, information on associated books or merchandising offers, announcement of competitions, all these (and there are many more) have to be produced and available to insert into a schedule at a moment's notice during the make-up of transmission tapes. If the channel is live-linked they need to be on-call by the station's presenter.

Complex animations and graphics can now be created on the desktop computer at very low cost and even if they cannot be shown live from the computer can be 'dumped' to videotape for playout.

Studios and Outside Broadcasts

A studio is a room with good sound qualities (proofing from external noise and internal reverberation); which offers adequate height for basic TV lighting, a smooth floor (so that cameras can be tracked); is easily accessible from the street (so that furniture and props can be moved in and out); has plenty of nearby rooms for use as control rooms, viewing rooms and artists changing rooms; and which has as much storage space as working space associated with it and close to it on the same level. It will need air-conditioning, a water supply, special electrical installations and the Fire Brigade will need to inspect and certify it.

Many buildings can supply a room which meets the basic requirements for adaptation into a studio. Avoid the upper floors of tower blocks where there are few lifts. Avoid buildings next to motorways and railway lines (or over railway tunnels). Try and keep away from the flightpaths of airports.

When the time comes to design a studio an expert should be called in. There are many sources of good advice and it is easy to make simple mistakes at the earliest stages which continue to have an impact for years to come.

Another area which requires expertise is the design and development of outside broadcast vehicles. The growing miniaturisation of equipment means that more functionality can now be crammed into less space, or the same space can have added functionality. Broadcast OB trucks which once handled 5/6 cameras and 2 VTRs are now handling 14 cameras and 8 VTRs. These mammoth beasts provide the network coverage of major sporting and entertainment events and their staff are highly skilled.

A local channel is most likely to look at a 'mini-mobile' - a concept first devised in Canada but now quite widespread. Up to 3 cameras and 2 VTRs plus a graphics computer and all their control equipment can be packed into a light van and assembled in a room on site. Alternatively using a pre-rigged van the equipment can be set up in the vehicle and used like a traditional outside broadcast truck or quickly de-rigged into a venue.

There are many dangers in adopting a do-it-yourself approach - especially if the productions tackled are over-ambitious. But strangely this is an area where less is at risk that in a studio. Capital is being spent largely on equipment which has a life apart from the OB/mobile and can be used elsewhere when it is not actually needed for the purpose for which it was acquired. The studio's investment is as much in buildings and fittings, which have little or no reuse value.

It is thus no surprise to find that many people looking at local television are considering using location production techniques, and are not planning on building a studio.

A studio will, one day, be required. This will be a small presentation, news and sound recording studio. Used for camera links, simple interviews and recording voice-over commentaries for programmes, when the channel schedule has grown this facility will be in constant demand. This investment will also probably be associated with a 'play-out' facility (built into it or controlled from it) and thus offer the channel the opportunity to link the output live at the times of day when a programme schedule is being transmitted for the first time.

Advertising Transmissions

The sophisticated advertising play-out systems used by the terrestrial broadcasters employ several play-out VTRs, with each ad on its own cartridge, loaded and unloaded by robots, and scheduled by a central computer running a schedule linked to the advertisement billing system.

When a cable network is planning to show adverts to be inserted into locally available advertising slots ('avails') on satellite delivered channels it is likely that some of the functionality of these highly sophisticated systems will be required. There are specialist playout systems for cable head-ends.

Advertising on a single local channel does not require this sophistication, but if it is in use for the 'avails' it might well be employed. It will be more usual for adverts for insertion into local channels to be inserted from specifically compiled tapes when the day's transmission tapes are made up, or from a made-up adverts videotape roll, with each advert block pre-edited, if live presentation is used.

Controlling the Costs

There is no escape from the fact that television production hardware costs money. It must be either hired or bought, and a channel with a continuing output is not likely to regard hire as a desirable route.

Even though equipment rapidly goes out of date in a sustained local channel operation all equipment available will be used well past the time at which others will have thrown it out.

Costs will be major influencing factor on the choice of the equipment with which the channel will start operations. It thus becomes important for a medium term view to be taken on the channel's growth, trying to ensure that any equipment acquired early on will retain compatibility with equipment which comes later. Second-hand equipment may have to be considered, and there are some very good second-hand bargains around because someone somewhere in the TV industry is always replacing or upgrading their investment.

It is unlikely that (at early 1994 prices) very much can be done with less than about £60,000 of capital expenditure - assuming Hi8 or SVHS standards. This would allow for a photo-videotext system, computer graphics, two location camcorder kits, 2-machine video editing, and the support equipment necessary to undertake documentary style production.

A mini-mobile with 3 cameras and VTRs of its own will probably come close to £100,000, but could be in use on a new location every day, and sometimes twice a day. This scale of production would require a second edit system (absolute minimum £10,000), and the transmission complexity may have increased so that there will be additional central needs.

Creating opportunities like a videobox, fixed contribution links to/from other sites in the cable network (perhaps a high street shop, football ground or the local council chamber) would all be additional. If these remote sites are to have their own remote unattended camera, microphones and communications, such as a local newspaper newsroom which could be providing frequent input, there will be remote control equipment costs to add as well.

Support

The second key issue with the technology of production is the quality of support required.

The simplest channel, showing previously made-up transmission tapes of acquired programmes produced in a documentary style, probably only has graphics computers and simple 2-machine editing. Its technical support needs could be provided by head-end operational management or an on-call maintenance contract with a local professional video dealer. Even better might be to get his engineer involved as a volunteer !

The most vulnerable elements in the simple production or post production system are the editing and graphics equipment. A simple training course for a member of the channel's staff (or even a volunteer who can be relied upon) may be all that is needed to ensure that the set-up and standards can be maintained. There will be the need for regular checks on the wear of videotape player heads and the supply of replacement machines while servicing takes place, but this will be infrequent and can be handled through ad hoc arrangements with supplying dealers.

As equipment standards rise, the variety of equipment grows and the amount of equipment in use increases some on-going technical support will be needed. Many channels will find that a combination of staff training for front-line fault-finding by their key suppliers (Sony, Panasonic and JVC all offer such training) together with maintenance contracts with their supplying dealers, will prove adequate. Operational staff such as video editors and sound engineers will usually be quite capable of handling first line maintenance and diagnostic work.

Once a programme attempts live programming the demands again increase. Live programmes have a tradition of generating unexpected problems and an experienced engineer with full familiarity with all the systems in use is the only way of providing a guarantee that the unexpected can be handled. When you plan for a mini-mobile tackling a live OB, linked through a studio and with adverts inserted from a presentation area, the complexity has once again risen and some level of competent engineering support will be needed at each of the sites involved. One of those engineers should have had a good experience of problem-solving all those systems. It would be normal to have a qualified engineer permanently allocated to a 'mini-mobile' and if it is working 7 days a week he would need a less well qualified, but competent assistant.

The channel which takes regular completed programming from contributing colleges or other local organisations will find that its technical staff become a local source of expertise. It will be hard to escape the fact that the channel's engineer will be called on to resolve every problem on anything technical which is affecting production which might one day be a local channel programme.

It will be one small penalty for being the recognised television leader in the community.

CHAPTER 12

A GROWING INDUSTRY

As the cable industry grows what influences will shape local television ?

The local television debate has been thrown into sharp relief by the development during 1993 of the 'super-franchise'. The larger cable holding companies have created groupings of neighbouring franchises, often with contiguous boundaries, and are developing them as single operations. Where boundaries are not contiguous licences have been sought from Oftel to allow interconnects to be built, and interconnects are also being built between neighbouring operations owned by different holding groups, sometimes across intervening, unfranchised territory.

Channel One London

The London Interconnect awarded Channel One Ltd, owned by Associated Newspapers and SelecTV, the contract for Channel One London in November 1993 ahead of bids from the Mirror Group and London News Network, the Carlton TV and London Weekend joint company.

The Channel One proposal is to start a cable exclusive channel of 'local' news, topical and entertainment programming for London and then to export the concept, the experience and no doubt much of the programming to other franchise areas. The eventual aim is to maximise the audience by providing a national/regional service using the cable industry's own interconnects. Once networks are in place, forming a 'super-highway', it will be relatively simple to undertake a national distribution to cable systems over the cable industry's own circuits and provide the core material for a 'local' network. The national press were quick to spot that this idea is not far removed from the terrestrial Channel 5 concept promoted in its application to the ITC by the consortium headed by Thames TV in 1992. The pressure for the ITC to re-advertise Channel 5 is being maintained. Pearson (owners of Thames) Time Warner (associated with the first bid) and MAI (owner of Meridian and Anglia) have come together with a new technical plan. This pressure for a city based TV concept, together with the 'super-franchise' as a means of delivering the concept more cheaply, although reaching a smaller audience, raises the issue of what is 'local' ?

In the original context of cable franchise applications 'local' meant within the franchise, even though links between neighbouring franchises were not discounted and were accepted in many later applications as we have seen in earlier chapters. It can be argued that this is the interpretation intended by Parliament as the only real discussion was during the passage of the 1984 Act. A London-wide channel is an excellent concept in a city which has retained little local identity in its centre and which has high commuter and entertainment traffic, giving people a sense of belonging to the overall city rather than their own locality. Those who feel 'local' to an area are generally the exception, often older, but they also include those who may be active in the voluntary sector or political life.

The existence of an interconnect channel does not invalidate the concept of more local services, nor does it replace any franchise obligation there may be to provide such services. It is those who are active in the life of the community who will remember the promises and keep up the pressure for them to be realised.

Outside London services on interconnects become less valid for the same reasons of identity. There is usually a far clearer understanding of what 'local' actually means. Distinctions in London between Barnet and Bromley are easy to find and it is impossible to conceive that Solihull and Sandwell would both accept the same channel as 'local' - such a channel might be suggested by the planned Birmingham/Black Country interconnect. A Preston and Liverpool channel is not likely to find favour in either area and Cable Northwest already recognises this fact by running two separate photo-videotext services.

A Yorkshire interconnect linking Bradford, Leeds, Huddersfield and Sheffield (as well as many other towns in between) would have similar problems if it attempted to identify itself as a 'local' channel. Indeed in West Yorkshire the creation of large metropolitan local authorities 20 years ago is still presenting problems for local government. Kirklees was formed from 18 original councils and each one of these areas still fiercely treasures its own identity. What chance of creating a loyal audience would a 'national/regional' channel have in these areas if it pretended to be local.

In all these cases an interconnect channel would be seen by the audience as 'regional' rather than 'local'. It will be competing with the long established and well equipped BBC and Channel 3 broadcasters who already provide regional services. Any sense of uniqueness which might come through 'locality' would be lost.

Third Party Channels

The arrival of major companies in the media to create and run a cable exclusive 'third party' channel (even if they are regional rather than local) is nonetheless very welcome. Also welcome is the recognition from Associated Newspapers and SelecTV that Channel One London is a long term project and will not quickly reach profitability.

The contract will give encouragement to other newspaper groups across the country who would like to look at local television as an extra 'leg' to their business. For many this will be their first real examination of an electronic publishing business opportunity. Like any significant business they will feel aggrieved if their approaches to be involved in such services are rebuffed by the cable companies.

The time is coming when cable companies will have to recognise that they may be unable to dominate their own networks and monopolise local TV services.

It is worth examining the cable companies' ITC licences. Every licence contains clauses about *"fair and effective competition in the provision of diffusion services and services connected with them"*.

The ITC has powers to direct that appropriate action is taken to ensure that there is competition. The ITC also has the powers to apply a Statement of Local Services to the licence, requiring operators to take note of promises made at the time of franchising and as amended in discussion with The Cable Authority at that time. If a local service was promised and is not being provided, the willing 'third party' prepared to fund a channel which meets some or all of the cable company's obligations at no cost to the cable company, may be very welcome.

If the price of such a channel is the advertising on that channel it could be a cheap price to pay. Contracts could ensure that the 'third party' is committed to a service level with which the ITC are happy, and possibly the cable company will pay a small per subscriber fee to help get the new operation under way. The deal may even allow them to take a carriage fee when the operation is eventually profitable. It seems a logical approach.

If the business of local television is to prosper cable companies have to be involved in the process of getting a local channel into existence. They control the 'gateway' to the audience and, ultimately, they are the ones who stand to gain the most.

My belief in the value of the 'third party' or joint venture approaches is that the emphasis in the cable business has changed. Telephone services are now the key to profitability, and the expectation of integrated TV and telephony technology during the 1990s which will endow telephony with the key characteristics of TV (and vice versa), is fuelling the growth of the networks.

Managing the multi-dimensional businesses that already exist is a major task. The complexity of telephony puts demands on the traditional cable TV manager, and the diversity and confusion of the TV world places demands on the traditional telephony manager. For whatever reasons neither is eager to add to the stresses of his job, nor add to the cash demands of his young business, nor add a further complication in his contact with a local culture which (it must be said) he often finds resistant to his TV wares.

The 'third party' route could be an easy way out of the dilemma. Someone else takes on the job. Someone else sells local advertising. Does that really matter just as long as the quality of the channel is such that it acts as an attraction to potential subscribers? A 1% lift on a subscriber base of 30,000 is worth over £100,000 per annum in turnover if telephony and premium pay TV services are taken as well as basic cable. The better the penetration the more of that added turnover is operating profit, the more of that turnover is telephony the better the ultimate margins.

If someone else has funded the subscriber incentives which create that growth, and have made their own profits into the bargain, everyone should be happy. If at the same time it is a service which enhances the cable company culturally and socially it must be a good thing to do.

Raising £100,000 in advertising turnover will have quite significant costs. The greater the turnover the higher the initial investment which must be made to create it. The margins of the small advertising business are unlikely to exceed those of supplying CATV and telephony to the same turnover and it is hard work to get it, every year, unlike revenue from loyal subscribers which eventually flows virtually automatically. It is a specialist game, leave it to the experts.

Local TV Advertising

Advertising lies at the heart of a commercial local television operation, whether the channel is operated by the cable company on its own or in partnership with another local advertising vehicle (probably a newspaper), or whether it is run wholly by a third party.

Co-operation between local newspapers and local cable television can be seen in an evolving way which most obviously revolves around advertising. Sharing advertising resources and revenue will be the key issue. This is not a problem faced in the other possible organisational ideas, but then the advertising market takes longer to penetrate.

Unlike newspaper advertising television advertising has always been heavily regulated, and structured as a scarce commodity commanding premium prices. It also carries a high production cost. This situation is eroding as TV channels proliferate, some offering niche audiences (e.g. music, children, sports fans). Airtime prices are dropping on general entertainment channels, Channel 4 offers a national TV advertising vehicle, and new satellite delivered channels come into the national market quite aggressively. It can never be divorced from the up-front costs in producing TV ads, but these are dropping too as the power-base of the ad agencies diminishes. However there is a limit to how low prices can go while still producing an effective TV advert.

Television advertising still faces effective regulation enforced by the ITC on content, and on airtime. No more than 9 minutes of advertising can be scheduled in any one hour on a local channel, plus no more than 3 minutes of TV shopping. The ITC is prepared to waive this rule for a local cable channel, particularly if there is the intention to provide a channel dedicated to local advertising.

For most businesses in the local economy the idea of TV advertising is a new one. They will be interested in reach and penetration, and in the shape of any deal which is offered. They will also be more susceptible than the larger national and regional advertisers to the emotional opportunity offered by local advertising exposure. In the USA and Canada local advert slots are noted for the advertising which show the owners of garages and stores promoting their own wares. Arthur Daley would be in his element.

Even if local businesses are genuinely interested in buying airtime there remains the relatively high cost of TV ad production. On the positive side adverts are always in colour, include sound, and scheduling can be very flexible.

It is easy for a local channel to under-estimate the difficulty of selling a local market to a national advertiser. The prospect of dealing with numerous carriers to reach a small percentage of the total TV market deters ad agencies from taking this opportunity seriously although we can anticipate that, one day, a new generation of agency airtime buyer will specialise in the niche and local opportunities which cable will offer.

The accumulation of all these factors often pushes the business plan for a local channel into looking at graphics production for adverts (including photo-videotext) as an alternative rate-card to full motion video production. Where local 'avails' are sold the range and variety of the airtime package which can be offered will be far greater and the advertising rates will be commensurately higher - than if only a local channel is available. The difficulty of achieving a quick and significant penetration of the market which justifies the significant capital costs involved in setting up to transmit 'avails', without the support of an external selling organisation, has been amply demonstrated by the experiences of all those cable systems which have tried to sell advertising to date.

In view of the developing trends and despite the fact that they are maintained within tight regulations, I would question whether revenue generation through advertising should be the key reason for a relationship between a newspaper and a local TV operation - even allowing for the potential strengths that appear to lie in the relationship. It would be an appropriate defensive stance for a newspaper to have some control of local TV advertising to protect the market share held by its core business. Whether they would allow the net new revenue generated (allowing for loss of volume on their traditional vehicle) to be significant is questionable, and inevitably the cable company will begin to believe it could do better on its own - whether that assumption is justifiable or not.

A cable company can accept and encourage local advertising competition if it wants to. Such competition is eventually inevitable as cable companies seek to maximise revenue from their networks without increasing their own costs. Advertising without editorial material to justify it in the audiences view is likely to be a weak service.

Avail Advertising

Experience of building a local advertising market started in the late 1980s when the industry still had fewer than 100,000 customers connected to broadband systems. Local promotional and advertising channels were launched in Westminster, Windsor and Croydon, and local channels in other parts of the country looked at accepting advertising.

In 1988 it was decided to launch the Cable Advertising Network and one advert at least was actually transmitted by all the franchises subscribing to the 'network'. However attracting ads was next to impossible - the industry offered little in terms of reach, identifiable niches, or absolute numbers and the network was more of a name than a reality. Although the 1990 Broadcasting Act called a halt to the pure advertising channels which could be run under the terms of the 1984 Cable and Broadcasting Act, as they were brought into the same regulatory structure as all other TV channels, the ITC will waive this rule for a specific local advertising channel.

The surge of new satellite channels which became available following the launch of Sky in February 1989 and BSB in 1990 brought another new feature into the cable market. The ad 'avail' - the optional regular slot into which local advertising could be inserted. At the end of 1993 the number of satellite distributed channels offering 'avails' of 2 minutes in each hour to operators had grown to eleven.

First to launch a service taking advantage of this availability was United Artists London South. In early 1993 they began selling the avail airtime across the whole of London through the London Interconnect. Birmingham followed suit and Cable Northwest, based in Preston followed in autumn 1993.

Cable Northwest has created a complete advertising department, with sales, creative and production teams, production and post-production resources, and sophisticated ad insertion equipment in the head-end. While creating a new advertising opportunity to local companies the approach is much more to co-operate with existing advertisers than to compete. A package which could include regional TV, newspaper, local radio or direct mail as well as local TV is seen as being the most effective use of the medium for many potential advertisers. Collaboration offers benefits to all, outright competition will cost everyone something but, as the cable staff are well aware, their parent company's pockets are deep and the revenue from the airtime they can sell will grow as the network grows. They feel that they are in a strong position to compete if that is what the other local media want to do.

Avails and Local Channels

These ad 'avail' operations are local television, but the way they will relate to local channels in the future is unclear. Their purpose is to generate revenue and profits from an advertising opportunity. They can sell airtime on local channels as well as on avails, and to avoid duplication of effort will do so. But what if a local channel is a 'third party' channel ? Will two advertising sales operation exist, each offering space on different cable channels in the same local area ?

The prospect of competition between a cable company selling space on eleven or more channels, and able to design a sophisticated campaign at highly targeted audiences, and a local channel with its own airtime seems an unequal battle. That the two should be one is clearly to their longer term mutual benefit. It increases the range of opportunities which a single sales operation can promote and sell and promises the local channel access to markets it might not otherwise be able to reach.

Taking advantage of the 'avail' opportunity is however likely to be at least as costly as setting up a local channel, and if a local channel already exists it is possible that an extension of its staff and resources may be the most cost effective route to realising a joint opportunity.

This opens the way for the cable operator to attract a 'third party' (or more than one) into his market to undertake the whole local origination operation, including adverts, in return for a 'licence' fee. It may be more attractive to enter into a joint venture, contribute some equity and access to airtime, in return for a carriage fee and of course the appropriate equity share of eventual profits. The imponderable saving from either of these approaches is the management time which would be consumed in setting up and running an in-house operation. It is also worth remembering the "fair and effective competition" clauses in the cable licence.

This finally leads us to the question of the business of local television. There is a wider advertising revenue opportunity than a single local channel; there is the prospect of using that revenue to create a local channel which appeals to subscribers and improves local relationships, sales, retention and penetration; there is the possibility of finding local collaborators; but what is the shape of the business plan ?

Does it stand up as a viable business in its own right when the common belief is that local television is a pure cost and should be avoided for that reason alone ?

CHAPTER 13

THE BUSINESS OF LOCAL TELEVISION

Commercial fiction or genuine business opportunity - can local TV can ever be profitable ?

It is in the fundamental nature of a cable operator to regard programming as a cost. This point was missed when the debate on the broadband cable opportunity started in the early 1980s. Even when the operator can identify direct revenue from carrying the service, the underlying channel or programming is a cost which he will assess on its ability to generate profits in the same way that he assesses any other aspect of his business. When he cannot see revenue, as is usually the case when he starts to fund local television programming, the resulting questions create tensions in his business.

Those who have to manage cable TV in the UK are well acquainted with such tensions. They have to face the constantly changing availability of TV channels and the vastly different world of telephony. The parameters which the customer instinctively applies to these two increasingly converging technologies are so different that building a common service identity whilst maintaining a judicious separation of the services has so far eluded them all. Most feel it important that programming issues, which can become contentious, do not rebound upon the telephony operation. It could be something of a tightrope act.

This viewpoint has been expressed by many of the telephony led cable operations and franchise managers with experience of cable in the USA are usually happy to subscribe to it. Their experience of access TV suggests to them that local television is necessarily of poor quality, will attract bad publicity and thus has a very real potential to impact on their credibility as high quality telephone companies.

There are also issues about the development of the interactive services which will be possible once integration of the technologies is complete. These will again change the perception of the network company in the eyes of its subscribers. I cannot see the highly respectable publicly quoted companies which own cable systems wanting to be seen as purveyors of socially undesirable interactive materials any more than many of them are willing to give any sales emphasis to the present soft porn TV channels.

This analysis is all very well given today's knowledge. It serves to highlight the real lack of considered data about the economic viability of local television. The fact is that no-one was able to predict either technology evolution or its likely impact at the time the industry's foundations were laid in the early 1980s. There was no effective challenge to the 'market forces' arguments which drove the Cable and Broadcasting Act 1984 and the Broadcasting Act 1990 through Parliament. The economic effects of that policy are one thing, the missed social opportunities are quite another.

If what we know today had been identifiable during the later 1980s it is possible that mechanisms could have been put into place to ensure that provision of local services was a priority. Specific terms in licences and innovative funding mechanisms could have helped the development of an important local opportunity.

Local Television - The Business Environment

Cable companies are building a local distribution system - a communications network - but they face a credibility issue. What they are doing is not universally perceived as either necessary or desirable. This attitude is very prevalent amongst residential customers. The research reported in Chapter 6 confirms what many believe from reported experience of cable salesmen.

But cable companies have almost unlimited access to airtime for advertising sales - why cannot they use this as a fund-raising vehicle for a local channel. Every cable franchise business plan assumed that in the medium to longer term revenue will be generated by selling space on all their networks, whether it is time/capacity (telephone calls or advertising slots), right of access (telephone rentals or subscriber TV fees), or leased/managed capacity (sale of telecoms bandwidth or carriage fees paid by TV channels). In each of these examples there has been rapid progress on the telephony element and rather less innovation and progress on the TV. The cable companies have fallen into a difficult trap, they are paying wholesale prices for the right to package multiple channels which offer them little distinctive or unique in the market.

The emphasis has been on the cable operator paying the channel operator to carry his material - a situation which the copyright legislation encourages and the latest regulatory rules in the USA reinforce. This fosters traditional broadcaster approaches to channel management and programme development and discourages innovation. While a range of these channels is necessary on the cable systems there is within the competitive TV environment a striking innovation gap.

At one level this is most clearly expressed by research indicating that local television is becoming a leading candidate to attract new audiences. It is now evident that this gap will only be filled by the adoption of new management approaches - to programming, development of low-cost production techniques and, most important of all, developing new sources of revenue.

The problem thus identified let us try to find the revenue to fund the programming opportunities we have already discussed. Given growth into profit it will be possible for a cable operator to take a fee for carrying the local channel. If he runs the channel himself it will be 100% of the profits. If not it may be a licence or carriage fee from the channel operator.

The local business opportunity may be more attractive for a third party than a cable company. Shedding the risks and acepting the benefits may be a real chance for a cable for operator. A third party interested and capable of setting up a local channel and sustaining an effective local service brings benefits to everyone.

The Third Party Channel Provider

Under the 1990 Broadcasting Act almost anyone can take a Licensable Programme Service Licence from the ITC. This has been recognised by a small number of programme providers who can offer programmes under their own channel logostyle for inclusion in a cable operator's line-up without creating an obligation on the cable operator to accept editorial responsibility for the content shown to the public.

By the end of 1993 only two third party local channels had been established. In Blackburn Open Campus Television is run by Blackburn FE College and offers educational programming. In north London Hellenic TV transmits to Greek subscribers in Cable London's network. It is a premium channel with, in the autumn of 1993, some 800 subscribers.

These two channels conform to generally recognised criteria identifying the potential members of local television consortia, or idependent channel managers. The likeliest people to be involved in 'third party' channels are:

- Channel 3(regional ITV) companies
- Newspapers
- Colleges
- Ethnic groups

Channel 3 Companies

The first group that has looked at local channels and negotiated actively with cable companies to run them have been the regional Channel 3 companies. To date every negotiation between the two parties has fallen down.

Analysis of the reasons appears to indicate three main causes. First the attitude of the regional ITV company is defensive. It perceives local channels as a potential threat and by taking the lead in any area it tries to keep competitors for its TV revenue out of that local marketplace.

Secondly, in a market where it may take two to three years to build reasonable revenues the necessary investment in a channel to the standard that an ITV company would want to be identified with is sufficiently large to make them think twice about the proposition. As the emphasis in the regional companies is to look for new revenue more and more nationally and internationally a new business with a possibly difficult route to profitability, and part of that at the expense of mainstream revenue, is not an over-inviting idea.

The third factor is that the ITV companies see cable companies paying a per subscriber fee for mainstream channels and seek similar arrangements. Cable companies are 'margin sensitive' businesses and look at such subscriber based fee arrangements only if they can identify an additional benefit coming from a contract. With an ITV company it is almost impossible to identify such benefits.

Local Newspapers

By contrast local newspapers have something very real to offer - advertising their involvement in their own pages, and thus advertising the cable company and its range of services. Newspapers are also more local than the regional ITV companies, despite efforts in recent years to make regional TV more local in its content. This local-ness comes through in two key ways.

Firstly a local newspaper has an extensive and well-established local news gathering organisation, equipped with the manpower and technology to handle the considerable editorial input required. Secondly it has an advertising sales organisation which is well established at a number of levels in the local economy - classifieds, run-of-page (display), whole/half pages, and advertising features with customised editorial. A local newspaper's business customer base will number thousands of contacts.

However it must be said, a newspaper's motives for getting involved in local television may also be defensive. Erosion of their revenue base by a new medium is not a new threat - local commercial radio in the 1970s presented the same situation. The radio threat did not materialise - new advertisers came into the market, and existing advertisers generally increased their spend to accommodate the new medium. The whole local advertising market grew. Newspapers hope that the same will apply again, but can they afford to take the risk ?

The local press uses a well-proven physical distribution system. A percentage of total revenues is raised directly from distribution, but the fact of the distribution attracts advertisers who provide the major proportion of revenue.

By implication there is threat to a newspaper if a cable operation provides or considers providing parallel local journalistic coverage or advertising in the local market. This threat grows as the cable build advances and the penetration achieved by the cable operation increases. Because it is a slow process and cable operators generally have the financial muscle to accept losses on local programme operations as part of their overall investment in the local market, the comparison with local radio of 15-20 years ago is not necessarily valid. Local newspapers look at reach and penetration just as intently as cable companies. Even when the numbers and the scale of the cable company's supporting operation are initially tiny compared with the depth of resource committed by the newspaper the growth of any threat will be intently watched.

It seems logical for a newspaper and its local cable company to seek to exploit their respective strengths for their mutual benefit, and to protect their existing interests. Just as with the Channel 3 companies no such relationship has lasted in UK cable. The only long standing venture is fairly relaxed. Croydon's 'Cable Today' still takes its local news from the Croydon Advertiser.

I believe that this failure to get closer together is only because there has not yet been a sufficient mass of cable subscribers in any one area to threaten a local newspaper's circulation or to attract local advertisers away from their traditional vehicles. The scales of cost in TV production have also been outside the experience of newspaper companies and the realisation that technology has now brought costs down takes time to be understood. Newspapers also, quite understandably, seek exclusivity for any service they offer so that they can retain a competitive edge in their own core business and compensate for the risks in a new market area. In some cases they are effective monopolies ready to use their monopoly powers. Exclusivity is not something that cable companies can offer without possibly attracting rulings from TV and fair trade regulators.

Colleges

The investment made by Blackburn College in their channel is significant. They have had the benefit of being the first to get underway and the encouragement of suppliers and funding bodies for their development work. Thirty hours a week is a significant programme output for 3 full time staff. Most of the programming is acquired, and there are well-established repeat patterns.

The motives for colleges showing an interest in local TV channels are quite often misguided. The financial case can be hard to make. Unquantifiable factors such as the perception of added value offered to the community by the college have to be taken into account. Can their TV network have a real effect on student numbers and help build a college's core funding ?

As networks grow established college cable channels will be able to reach outside their traditional geographical catchment areas and attract students, full-time and part-time, that would not have enrolled before.

The introduction of interactivity into cable will also, eventually, offer colleges new opportunities. Distance learning for full-time and part-time students is a proven effective learning method. The ability of a cable system to offer distance learning though interaction on live and recorded programmes, in real and deferred time, to niche audiences which could be as small as one person, could be attractive economically as well as being effective for the learner.

It is thus possible to accept the motives of colleges wishing to be involved even if a hard-nosed short-term business case is more difficult to identify. First in may not be the winner, but he stands a better chance.

Ethnic Groups

Many ethnic groups wish to make their own programming and run their own channels. As with colleges quite often the enthusiasm will exceed the abilities which are available to resource and manage such a project effectively.

The experience of Hellenic Television in north London is quite salutary. A small staff manages the schedule, tries to attract advertising from the community and, with volunteers to help, make local programmes. The channel takes in Greek TV from satellite and re-transmits it to provide most of its schedule. The localness has been relatively small. Resources are tight and the wide variety of skills required cannot be encompassed by the few staff who can be afforded on the revenue raised.

Hellenic has a loyal following but it also offers a ready invitation to VCR owners to record and distribute programming for neighbours and friends, regardless of the threat to the service which that implies.

Another ethnic channel experiment in south London allocated weekend hours to individual producers. Each producer was required to obtain his own ITC licence for a Licensable Programme Service before being allowed to transmit. This eliminated the need for Videotron, managing the channel, to take an editorial view of every contribution - a thankless task when a high percentage of the programming was not in English. Many of the producers were not local to the audience, although they attempted to keep their material relevant.

A better route to providing a local ethnic programme service may be the allocation of set hours on a general local channel. It can be open to accusations of creating media ghettoes such as have dogged broadcasters in the past, but it is flexible and allows the scale and scope of programming to be contained within the capabilities of an interest group.

The inevitable vagaries of delivery of material are easier to cope with and by forcing the producers to stay closer to their sources the importance of earning revenue is hard to avoid.

Advertising Funding

Concentration on the advertising issues and opportunities clouds the real debate. A mixed media market for local advertising is inevitable over 5-7 years. How existing advertising vehicles face up to the potential for competition from local television is to a large extent dependent on growth in local television itself - they must compete or complement each other editorially.

The objective of raising advertising revenue needs to be to fund the local TV channel. We have already discussed the dangers of being seen to be 'milking' the local market.

So how much money can be raised ? And how does it compare with the costs of running a local channel operation ?

Analyses of US experience suggest that an attributable advertising revenue of £8-10 per annum per household subscribing to a cable network is conservative, but realistic. The advertising operation in United Artists London South grossed close to £250,000 in 1992, serving a network growing from 30,000 to 40,000 customers (approx £6.25 per annum per household). It exploited one local advertising channel and advertising 'avails' on 6 further channels. Some estimates for gross revenue when the full range of 'ad avails' are exploited (now 11 channels) plus one commercial local channel, reach £18 per household per annum.

A realistic approach is to assume that gross revenue per household will grow in line with the cable network's increase in reach and penetration. The medium becomes proportionately more attractive as the system gets larger.

The total UK cable revenue potential for advertising on 'avails' and local channels on this basis looks something like the model in Table 1.

Even at the higher rates shown in this table the gross revenue would not become a significant share of national TV advertising revenue. Jonathan Davis's September 1991 report 'TV-UK' estimated that total national TV advertising revenue in the year 2000 would be £3.678 billion. The highest rate for advertising revenue in Table 1 produces a result which represents only 3% of this total.

This figure may seem modest but calculating it another way produces approximately similar figures.

The £3.678 billion estimate for gross national TV advertising revenue equates to approximately £56 per head of the UK population - a modest increase on today's figure. There are an estimated 22.5 million TV households - thus each household generates £162

Table 1: Gross National Ad Revenue Potential Per Annum (£m) for Cable Locally Inserted Advertising

		£ per household ad revenue per annum					
	Subscribers m	6	8.5	10	12.5	15	18
End 1994	1.12	6.72	9.52	11.20	14.00	16.80	20.16
End 1995	1.71	10.26	14.54	17.10	21.38	25.65	30.78
End 1996	2.38	14.28	20.23	23.80	29.75	35.70	42.84
End 1997	3.23	19.38	27.45	32.30	40.38	48.45	58.14
End 1998	4.30	25.80	36.55	43.00	53.75	64.50	77.40
End 1999	5.28	31.68	44.88	52.80	66.00	79.20	95.04
End 2000	6.12	36.72	52.02	61.20	76.50	91.80	110.16

Subscriber figures based on CTA estimate - Winter 1993

per annum in ad revenue, suggesting an average of 2.9 persons per household. At a population per cabled household of 3.2 people (as in 1992 estimates from the ITC) the gross TV revenue per household produces an average of £179 - an increase over the national average of £17 per annum. This 'premium' value of the cable subscriber is remarkably close to the figures in Table 1 above.

Accepting this increase in total ad revenue is not unreasonable. It would come from existing local advertising revenue redirected to TV, plus growth in the local advertising market, as well as more varied purchase of local airtime by national and regional adverisers filling 'gaps' in other media coverage. The assumption is not unreasonable. The sums are well within what is known today about trends in the ad market. It also allows for the fact that TV advertising in the multi-channel environment is reducing advertisement rates generally even though there is better targeting.

As a national revenue estimate these year 2000 figures certainly seem achievable. The impact will not be national. It will be concentrated on cabled areas where the households connected to cable by then will account for about 30% of all TV households. Shared across the national cable market the local advertising revenue is not likely to be that significant (the gross revenue for the cable industry by then will probably exceed £2bn) but it does relate to local TV and is currently a unique cable opportunity. A significant diversion of regional or national TV spend is not anticipated although the growth of the 'super-franchise's with 1m or so homes, suggests that cable may also be competing regionally with Channel 3 once these large networks are completed.

A Local Market

While small in the context of a national TV advertising it is important to recognise that the effect of local TV advertising will be felt disproportionately in some local markets.

There has long been a debate in the local television industry about the size at which entering a local market is viable. Some early estimates placed the bridgehead at 10-15,000 subscribers and some at 35,000 or more. The general agreement today is that a figure in excess of 20,000 is much more realistic. How much this is subscribing to an approach which defers the need to invest in the local market is a matter of supposition.

The experience of the few who have tackled local advertising seriously points to the need to enter the market early and to position the local TV advert as a high value product. The channel must be prepared to discount initially both to attract new advertisers into the market and to counter any effects created by the limited reach of a cable system which is still building.

It is safe to assume that while a percentage of the growing revenue will come from new advertisers an important element could be diverted from existing local and regional advertisers - currently using newspapers, radio and regional TV - but this diversion is unlikely to be significant until the new medium demonstrates results.

Thus getting into the volume market takes time, and investing early means that the results come earlier than they otherwise would. Researching the reach and effectiveness of the advertising carried is important. The data will need to be relevant to the local market so it will need to be borne as a cost by the local channel and any ad avails operating budget.

Given a considered growth rate for both the cable network and the advertising rates the following theoretical example in Table 2 could provide a core 'top-down' advertising revenue model. The penetration figures on which the sums are so dependent are open to argument. The national penetration at the end of 1993 was 21.9% although a number of franchises were succeeding in consistently reporting figures in excess of 30%.

The year 5 revenue of £1.28m demonstrated in Table 2 is certainly sufficient to carry the costs of a local television channel built on one of the models developed earlier in this book, and the additional costs of inserting adverts into local 'avails' on national channels. The underlying question is whether such revenue could

Table 2: Example franchise - growth of local TV ad revenue

	Year 1	Year 2	Year 3	Year 4	Year5
Homes in franchise	200000				
Homes built	20000	75000	130000	185000	200000
Penetration	20.00%	24.00%	28.00%	32.00%	35.00%
Starting early					
Ad revenue per household	4	6	10	14	18
Ad revenue (£000)	16	108	364	828.8	1260
Starting later					
Ad revenue per household			6	10	14
Ad revenue (£000)			218.4	592	980
Total starting early	2576.8				
Total starting later	1790.4				

be reached or sustained in the absence of local TV programming to build local relevance and local loyalties.

The figures give good reason to believe that a local TV business is feasible. TV production economics (as with most media businesses) have a relatively high fixed cost, and a nil or negligible marginal cost per additional household served. As the cable network grows, bringing new subscribers on line, and increases its penetration, there is no additional cost to the TV channel of this growth, but its revenue potential increases.

If the two (programming and advertising) can be built, and work together, both the channel and the cable operator take unique benefits from their presence. A local channel structured and funded on this kind of model can stand alone as a business and be profitable. It will offer the cable company a local service it can promote in its sales package, can use as a customer loyalty and retention vehicle, and can use to promote its range of other services such as telephony, pay-per-view etc without any additional overhead burdens.

This service does not come entirely free and a cable company wanting to take advantage of this opportunity itself has to adapt its corporate thinking to accomodate a kind of business which is poles apart from its traditional sector. Firstly the start-up, although the capital requirements are small in the total context of a cable investment, the start-up requires a disproportionate amount of management effort - and it is specific expert management that is required.

Then there is a different approach to how capital is accounted for. The long term capital write-downs that cable and telephony can be secure with are not possible in TV production. A 3-5 year write-down on much of the television equipment is too long. Intensive use wears out production kit and the volatility of quality standards requires that the development of new technqiues is actively followed. Depreciation and re-investment must go hand-in-hand.

It must be recognised that in its early years this operation will not cover its costs. A new cable operation is full of different examples of this kind of investment but the numbers of subscribers drive the revenue directly. With advertising the drive is indirect and there is a delay while the investment in sales effort takes effect.

Finally, as its network grows, the UK cable company is likely to begin to identify itself more as a telephone company offering TV services than as the traditional North American model of a pure cable company. This decreases the emphasis on local TV, culturally places more emphasis on tightly controlled overheads, and reduces any sense of necessity to get into meeting the opportunity for developing a local advertising medium on TV.

Other Local Revenue Generators

There are a range of possible secondary revenue generators where a local channel and a cable company can work together to their mutual benefit - regardless of whether the local channel is a standalone 'third party' operator or owned by the cable company.

The promotion of premium telephone lines on TV offers a potentially valuable additional source of revenue. ICSTIS (the Independent Committee for the Supervision of Standards of Telephone Information Services) has a Code of Practice which governs such services and the ITC Programme Code has very specific requirements governing the on screen promotion of premium rate phone services. Revenue is split between the telephone company, the premium rate service provider and the information provider. The cable company can seek the agreement of the regulators to be the telephone company. The information provider could be the channel or the cable company. The service producer could be either, given the necessary investment. Thus this opportunity allows a cable operator to take a larger proportion of the revenue than would be possible by using national services. However there is no evidence yet that totally local premium telephone services are viable.

The local channel could provide local TV shopping segments, using the basic model of QVC(now available on satellite). Stimulating the local economy in this way could add value to the perception of cable and local TV in the minds of local businesses, building relationships which may have spin-off value to both (for example through increased awareness of telephone services and increased TV programme sponsorship opportunities).

Competition or Collaboration

Turning the cable operator's power as an absolute gateway to his network into revenue is an issue which has not yet been fully explored in any public discussion or debate. A cable operator as a channel operator, covering costs and taking profit from the advertising revenue created, will be attractive to some cable companies but not all. It is the local television option on which most focus has been placed and it is the one which conforms to both past UK experience and experience elsewhere in the world.

But how much more attractive is the idea of allowing third parties onto the network to provide commercial local channels and eventually other local services? There is no reason why there should be only one local channel and no reason why there could not be active competition.

The first, and possibly most convincing, argument in its favour is that the cable operator stands to gain from an extended range of local services (assuming a base level of quality) at no risk to his business. He makes no investment, faces no close down costs if any services fail commercially, can publicly distance himself if necessary and take credit if the opportunity arises. Access to the network (and thus to the subscribers) can be bid for by competing channels. They may propose a fee linked to the advertising revenue generated (e.g. a percentage within fixed bands) so that success is shared. There may be time limits and performance milestones which can be agreed, and various incentives may be offered for a period to help a new channel get established.

A profitable franchise area, subject area, or specific service could be 'auctioned' - similar to the way the ITC auctioned the Channel 3 franchises in 1991/92. Maybe one local channel wishes to commit to a live or as live schedule, and another is looking at providing documentary, information and education programming for a subscription premium (shared with the cable operator).

The availability of such options will benefit the overall cable service on any network - and the cable company bears no risks and virtually no costs (capital or revenue) in taking this approach. Yet it is an approach which has not been thought through by the world of television, nor has any cable operator suggested to potential suppliers that this will be a future route to his network. It may not be long before some of the'super-franchises' begin to entertain these concepts.

It seems unlikely at present that any cable company will invest in a full TV operation of his own if he can attract a partner. As discussed earlier Channel 3 (ITV) companies have obvious potential but will tend to look at services for cable as a defensive strategy and are likely to be more attracted by interconnects because of the overlap with their existing service areas.

A local newspaper seems the most appropriate partner for a cable company, and also one of the most appropriate members of a local consortium for running a third party channel. Colleges form another group, though as already intimated their motives are different and it is doubtful what direct commercial value they bring to a channel.

Costs

We have debated revenue, and advertising revenue in particular during this chapter without considering the costs of running a local TV operation in any detail. They are of course equally important because the outcome, profitability, is the final deciding element for any investment.

One of the culturally difficult factors for a Channel 3 company looking at a local cable TV channel is that it finds it difficult to conceive of the three elements which are essential to a low-cost operation, and has no experience of them. These are:

- low-cost equipment (even throw-away equipment)
- volunteer effort
- direct 'door-knocking' advertising sales.

The Canadian experience suggests that a four-hour a day TV channel with 20 hours of computer rolling graphics, containing exclusively local programming, can be maintained on a cost of half-a-million dollars a year (approximately £300,000). The biggest cost element is staffing, but most manpower on programme-making still comes from volunteers, or local people paid on a freelance basis.

Running a news operation adds to this cost. Full daily hard news with reasonable moving image coverage from a relatively geographically constrained area will cost at least a further £300,000 per annum. The costs come through the combination of editorial and technical staff (likely to be at least 10-12 in number and as many as 20 with news) and the need for more robust professional production and editing equipment. Alliances with other local news operations can bring some of these costs down, but the essential elements of cost - TV staff and equipment - are unavoidable.

A Cost Model

Based on the experience of Canadian and UK systems to date (see Chapter 7 on working models) the following Table 3 provides a guide to the operating profit and loss in a local television channel relying on volunteer efforts and including a modest news operation.

Table 3: An Outline Local Channel cost model

	£000	
CHANNEL OVERHEAD		
Channel manager/NHI etc	45.00	
Admin/finance staff etc	50.00	
Rent etc	42.00	
Telephone and communications	25.00	
Cars (leased) and transport	31.36	
Equipment maintenance	35.00	
Routine consumables	15.00	
		243.36
PROGRAMME COSTS		
Staff salaries etc	180.00	
Freelance staff and exps	26.00	
Artists and facility fees	26.00	
Tape stock	37.44	
Programme acquisition	50.00	
		319.44
AD SALES (based on sales of £1.28m)		
Sales management	30.00	
Sales staff	110.00	
Commission	112.50	
Advertising/marketing	100.00	
		352.50
TOTAL COSTS		855.30
AD REVENUE		1280.00
TRADING PROFIT		424.70

In addition to the on-going costs illustrated there will also be capital costs. It is likely that a channel of this kind will require a capital spend of the order of £500,000 for TV production, editing and head-end equipment, and a further £150,000 for the ad avail insertion equipment. There will be a continuing need to update capital equipment requiring at least £50-75,000 each year.

The number of hours of transmission, of original local material and the percentage of acquisitions, will determine the final shape of the business plan. Anyone embarking on advertising, whether with 'avails' in addition to a channel or not, will need to look at how the revenue potential divides between the two kinds of service.

There is little doubt that the avail operation will subsidise the television programming. In the Table 3 the revenue, costs and profit could be assigned between the avails and the local channel. The lower cost and higher revenue opportunity presented by the 'avails' would subsidise the local channel. However the local programming increases the attractiveness of the whole local advertising package, the ready availability of production resources creates an additional revenue opportunity from producing adverts for local advertisers (or for hiding discounts) and it also reduces the likelihood of accusations that the cable company is there solely to 'milk' the local economy for the benefit of distant shareholders.

Taking the two advertising opportunities creates a viable standalone business. The cable company could be a member of a consortium, handle matters entirely on its own, or allow third parties to run the operation, recognising that his benefits would come from other directions. A channel such as that modelled in this chapter is not likely to produce final before tax profits much in excess of £100-150,000 per annum and within the total context of his longer term profitability such profits are negligible.

How much the market might grow by the introduction of competition is impossible to estimate. Introducing competition is something that will be undertaken cautiously because the potential for depressing local advertising rates to the disadvantage of everyone is very real.

It will be for each cable company to decide. As the owners of the networks there is no route to any local TV audience without their agreement. There is no eventual benefit without risk, but if cable operators would be prepared to take the third party route they could speed up the introduction of local television for their own benefit, meet obligations which their company acquired when the franchise was won, develop a small but risk free stream of revenue if there is a profit sharing agreement with the channel operator, and increase their local profile in a very positive way at nil or negligible cost in management time and capital expenditure.

Other Funding

Some structures for third party channels are also likely to be eligible for small elements of funding support from bodies such as the TECs (Training and Enterprise Councils), local authorities, development agencies and local development projects initiatives such as City Challenge. It is unlikely that some of these sources would regard a cable operator owned and managed channel as an appropriate destination for their funding.

There are additional commercial sources of funding. Generally programme sponsorship will be seen in a similar way to advertising by local companies, and it may be unwise to consider it separate. The additional revenue sources could include:

- merchandise - also valuable as 'giveaways'
- publications linked to programming
- revenue from premium rate telephone lines
- copies of tapes of programmes
- sales of programming to other operators or even the major networks.

Whilst none of these revenue streams is likely to be as important as advertising and sponsorship they do represent an additional opportunity.

CHAPTER 14

MOTIVATION AND SKILLS

A new economic focus for local TV

It will have become amply apparent that creating a local TV channel which stands a chance of being commercially viable will call for a wide variety of skills, not all of which will be readily thought about when a channel is considered. The other consideration is the motivation of those who have already, or will be expected to acquire, the skills which make the channel perform effectively over a long period of time.

It would not be difficult, based on the model in the preceding chapter, to argue that the key skills are those of selling advertising in the local market, although every programme maker will quickly counter that without watchable programmes a meaningful sales performance will not result. This debate is fruitless. The truth is that the range of skills and the motivation of the team involved in creating a local channel must be addressed as a whole from the outset. Sustaining the structures is just as critical. It must be able to cope with motivating new recruits, both staff and volunteer, if the local TV service is to be effective.

The sales team's skills are critical. They will encompass a wide range of otherwise disparate experience. A key role is selling in the local market at rates which attract the small business, particularly the retailer. This implies a service which compares and competes with the local newspaper. The experience of the local press is that a single sales representative might achieve £120-150,000 annual turnover. Local radio sells at higher equivalent rates for its reach but its airtime is restricted. Slightly higher turnover might be expected from a radio sales rep who is also more likely to be 'on the road' than the majority of the local newspaper's staff.

A 'photo-videotext' service backed by audio can be sold in a similar way to the local paper. The existing evidence (from the USA, Canada and the UK) suggests that it is likely to prove to be a fruitful market if tackled in the correct manner. It offers a new full colour

Table 4: Local Channel Skills Needs

FUNCTIONAL SKILLS		
SALES AND MARKETING		
	market research	sales management
	direct sales	tele-sales
	sales admin	credit control
ADVERTISING		
	commercials design	ad production
	account handling	
TEXT AND EDITORIAL		
	information gathering	compliance
	writing/editorial	PVT design
	PVT authoring	teletext authoring
	schedule promotion	technical support
TV PRODUCTION		
	TV production	operational skills
	channel promotion	programme acquisition
	commissioning	editorial skills
	copyright	compliance
COMMUNITY & TRAINING		
	community relations	on-camera skills
	training in the community	
	training for professionals	
MANAGEMENT SKILLS		
GENERAL MANAGEMENT		
	management skills	people handling skills
	intellectual property	employment law
	editorial adaptability	technology vision
	'warm' and 'tough'	product development
	regulator relationships	PR and press
	local ethnic 'politics'	health and safety
SECTOR AWARENESS		
	education	business
	voluntary sector	government
	TV and communications industry	
FUTURE VIEW		
	nature of interactivity	benefits of interaction
	mechanics of interactivity	

advertising service on TV at rates hitherto unavailable. It is an attractive proposition to bring to the market.

In many cases the sell will be 'emotional'. Even though local TV advertising has been available for more than a decade in many US and Canadian areas the attraction of fronting your own advert, whether on a still photograph or as a presenter in a short movie ad, seems unfailingly attractive to many local businessmen. There is only superficial anecdotal evidence of the effectiveness of such advertising but who worries - if the advertiser is happy so is the cable company.

Selling full motion TV advertising in the local market is a new business and although it calls on some of the experience of network selling the reality is closer to that of local radio. Such advertising is likely to be bought direct, rather than by agencies, and deals will probably be inclusive of production in the first instance. Once an advert is made the attraction of using it again and again, relatively cheaply, is also a unique attribute of a local channel. Such advertising selling does not compete with regional ITV companies except in a very small number of cases. The reach is different and the deals are very different.

The availability of movie ads in the local market will also offer a logical expansion of existing campaigns by national or regional companies, through re-cycling their national adverts or new locally produced adverts for local outlets exclusively for local use. Selling the national/regional and selling the local are wholly different markets requiring different skills and experience.

Sales management is also an issue. A local cable channel will require a very varied sales force possibly numbering more than a dozen in a larger areas. Amongst them will be direct sales experts, telephone sales staff, contract/account management staff and those who develop deals with other media (such as joint selling with newspapers and local radio) and inclusion in packages for national and regional campaigns. Sales managers will need to be on the ball and have some experience in all sectors (or be prepared to get it the hard way very quickly).

Training Needs

The ratio of sales staff to technical and production staff will depend on the policies the channel develops for commercial activity, acquisition of programming, for production and for use of volunteer effort.

The range of skills required in the channel is summed up by the chart. This identifies a number of major skills groups and a number of skills in each group. These are shown at a relatively broad level because there are clearly many individual skills entailed in achieving an overall level of competence in the operation as a whole.

There are also a number of issues which are difficult to address. A good advertising creative writer/director/producer is essential if local ads are to be produced, but any creative individual who demonstrates real ability will be in demand in the wider world. We have already seen how cable can act as a training ground for presenters and technical staff. It has been a well trodden route in the US and Canada and there have already been some notable transfers in the UK.

Providing the training and the working experience which makes skilled people desirable in the bigger markets is something that cable channels will simply have to accept. The trail from local to national is even more evident in journalism. Many of today's household names on television news and current affairs have trod the path from first experiences in local newspapers and radio stations. The implication for local television stations of this overhead is something which it is difficult to calculate. It seems ironic that the growth of local TV should be happening at a time when formal training in the television skills is becoming something which broadcasters (including the BBC) are finding it less and less easy to support. The possibility that local TV may become an (unfunded) trainer for the rest of the TV industry is very real and I do not see how that situation can be remedied easily.

This implies that a noticeable line in the budget of any local TV channel will be training. If a quality level is to be maintained when there is a queue of attractive offers for the staff (and the best volunteers) there will be on going skills development role. The opportunity for the individual to prove that he/she has something 'extra' will be one of the key motivations for joining a channel's team.

Motivating the Team

This all raises the issue of motivation. To some extent this has also been discussed in the chapter on volunteers. However there is an on-going debate about the social aspects of local television and it is relevant to introduce the issue of the 'social economy' when considering motivation.

The capitalist cable companies have shown themselves unwilling to invest in local television. The use of the word capitalist is deliberate and not at all intended as derogatory. The massive investment required can only be undertaken by governments or those with the ability to access large cash reserves (of their own or other institutions).

It will have become clear that the main arguments for creating local television channels are not ones which appeal to capitalist motivations. Creating returns for capital invested which exceed interest rates are what interest capitalist institutions. The motivations of those which pursue local television are different. We have seen how difficult it is currently to assess the impact on subscriber levels which a local service might have. Apart from profits this is the only other motivation which most UK cable companies seem likely to find persuasive.

However the social and local economic structure of the localities which cable company serves are different. The benefits of the new infrastructure the cable company builds will be accepted and welcomed in due course, although 'excess' profits (however they are defined and if they are ever generated) will lead to a degree of disenchantment (as witness the current dis-satisfaction with British Telecom). Ignoring the social issues, when there is evidence of increasing interest by society at large in these issues, is only done at the peril of losing public support.

The cable company which fails to listen and respond to local concerns about environmental disruption during the build offers a substantial piece of evidence to local people about its disinterest in them and their priorities. These are the future customers and they should not be ignored.

The Social Argument

This becomes pertinent to the potential for 'third party' channels. A channel with 'social economy' principles (rather than purely capitalist principles) can help to alleviate some of this potential public relations pressure. A company which is run on the lines of a co-operative (properly structured and incorporated) can be an immediate and obvious new employer, can openly and honestly state that advertising revenue is re-cycled in the local economy and that profits are similarly routed back into its own endeavours, not syphoned off to distant shareholders.

The issue relates strongly to volunteers because the motivation to communicate can then be reinforced by a social motivation which is deeper than simply a concern about the subject of the programming. The doubts that have been expressed about this concept of the local channel as a company with social objectives concern standards and professionalism.

Jim Brown (of Brown Baker Associates) is a research consultant who has developed a body of work on the social economy. He has defined, not without some debate from those working in the sector, a number of underlying differences between 'capitalism' and the 'social economy' which strike me as very pertinent to any debate on the purpose behind local television (and therefore the motivations of those involved).

He suggests that the capitalist philosophy is encapsulated by

- the principle of self-interest
- the pursuit of private wealth
- the right of freedom from constraint by government
- the natural hierarchy of man.

In contrast the core philosophy of the social economy can be defined by two important ingredients - "the distinction between material wealth and quality of life, and, the notion that we may act out of concern and compassion for other people and not just in pursuit of private gain".

There have been many successful examples of employee-owned companies and co-operatives working in the open market. They do not produce sub-standard goods any more than any other company, they can have managements which are just as good as any multi-national, and there are many such firms which work at the top end of the quality range in their spheres of activity.

The work of the Rochdale pioneers in the 18th century created a group of principles which are accepted in co-operatives worldwide. Jim Brown has modernised and simplified the language of these principles to help define a 'social' company:

- no restriction on ownership
- one-member, one vote
- limited return on shares
- fair distribution of profits
- commitment to education
- co-operation in areas of mutual interest.

These principles accord well with the generally expressed desires of volunteers in any walk of activity and offer them an appropriate opportunity to particpate in ways that other structures could not permit. This flexibility between the potential for volunteers and staff roles (whether full or part-time) is an important concept which would have been unthinkable only a few years ago.

The 'Social' Channel

The prospect that local companies founded on these principles could run local channels should not threaten any cable company. Indeed it may be that this route could be a preferred route for the creation of 'third party' companies. Some financial backing from the cable company and collaboration with colleges, newspapers and other supportive orgnaisations may be required - thus providing capital equipment and bringing in specific expertise - but this concept does not contradict anything in the Broadcasting Act.

Concerns over editorial control which are often expressed - both in local communities about cable companies holding the final say, and by cable companies concerned about a free atmosphere - are easily handled. The sense of 'democracy' which is implicit in an organisation structured on these principles actually provides greater safeguards with regard to observation of the ITC Codes, as all the members of the organisation have a vested interest in staying within the law. Their livelihood relies on it. Equally a structure can be agreed which ensures that 'unfair' treatment of the carrier network is avoided, without such an agreement providing constraints on fair comment.

In addition the cable company gets all the benefits it possibly could from the creation of a local television service, and its success. The cable company has the assurance that profits are re-cycled into skills, programmes and local activity which reflect credit on the network, and not into the hands of otherwise disinterested investors.

Drawing together the need for skills and the environment which can provide on-going motivation in this way provides a stark contrast with today's situation.

The few local channels that are running are almost exclusively funded by cable companies. Those cable companies with the largest subscriber numbers which are now generating revenue from advertising in the local market are not doing so whole-heartedly. The sums raised are limited partly because the investment in people is limited, although it must be said that it is still very early in the life of the industry. There has been no suggestion in any quarter that a viable local television service could be introduced by a cable operator and, in truth, a 'top down' organisation emulating a broadcaster and working within a large corporate hierarchy is unlikely ever to do so.

It would be an opportunity sadly lost if the cable companies did not now seek the 'social' collaborators who would venture with their time and effort to create local TV channels free of hierarchies and built 'bottom up' from a base of revenue fairly earned and skills effectively employed.

CHAPTER 15

THE FUTURE

How will local television play its part in the future TV environment ?

The future of local television in the short term seems uncertain. Cable companies have been reluctant to provide the substantial sums of money needed to generate local television and even more reluctant to spend the time taken to learn what running a TV channel is all about. There is even less interest in community television as a concept and despite the promises made at franchise time there is no enforcement action which can be taken to ensure that they meet this 'obligation'. For the enthusiast, whether a television professional or a community activist, the fact that the ITC (and the Cable Authority before it) have been given no effective powers to enforce franchise promises is a matter of sadness. It is worth remembering the ITAP report in 1982 made much of the potential value for programme and educational services on local cable television networks.

We do have to exempt the Canadian owned cable companies from this general criticism. Although Maclean Hunter (regrettably) extracted themselves from their franchises in the UK they showed one way forward. Their overall experience of the UK was not comfortable, despite rapidly becoming the leader in local TV. As the recession in Canada starved the parent company of cash so the decision were made to freeze building their UK networks. Stalled by the economic climate half a world away the East Lancashire franchise failed to reach a level of penetration which any MSO would regard as acceptable. If it had then maybe the message would have been different.

Thankfully Leicester Communications (owned by Sasktel and Fundy - both Canadian) and Telecential (whose key shareholders are CUC Cablevision and Telus Corporation - also both Canadian) are continuing the important Canadian influence.

Whatever the enthusiasm the commercial fact of life is that the viability of local TV has yet to be proved. The enthusiasm of the early 80s gurus did not bear fruit and it is still too early for full proof to be evident from the few advertising operations which are underway.

The long term for local television looks rosy. Television is fragmenting and segmenting. There is continuing growth of opportunity for new TV channels - more than 20 English language channels have launched or are about to launch in the two years since this book was conceived (1992). How many of them will survive in their present form is a separate question.

Meanwhile the evolution of technology is creating an environment in which the capacity of systems to carry new channels will become almost unlimited and telephone networks will be able to carry TV on demand. Fragmentation is being accompanied by segmentation. Channels targeted at niche audiences are growing more common. New channels rely on a mix of repeats, acquired programming and limited, highly promoted origination made at a low cost, for their schedules. It is a sad commentary that when the same programmes are re-cycled over a range of channels the identity of the channels begins to blur and any sense of uniqueness is lost. Technology also sets the scene for an extension of pay-television in a cable structure - through pay-per-view, on-demand or a near-on-demand basis. This all looks like a good environment for a unique local channel.

Then there is the new agenda of interactivity. Interactive television is likely to be a growing and continuing phenomenon during the 1990s, and interactive information services sourced through the TV screen will become a reality. Such services will offer random access, services will be on-demand and the apprearance can be individually customised. The change that this will create for users, carriers (cable companies and telecommunications operators), the regulators (ITC and Oftel), and for the wide range of information providers is potentially dramatic.

Interactive local services will become a reality. The opportunities for challenging well established conventions are legion. I remember explaining to local councillors in a north London borough how interactive television could offer their constituents a greater share in local democracy. Planning meetings could be reported with a TV programme which, through pre-recorded inserts, looked at all the details of each application. Following each insert a vote could be taken from the viewers. The concept is technically feasible and to many people the idea looks socially valuable and seems innocuous.

But it is an idea which daunts traditional elected local councillors when they hear that each constituent could be enabled to vote on each application, and the votes from each constituent area can easily be fed to the councillors for that area present in the council meeting and themselves about to vote. Technically it is only an opinion poll - they have to take the decisions !

This may be an extreme example, but it takes little understanding of how conventional organisations work to recognise the huge implications of such capabilities. The ability of the technology to create an 'underclass', effectively disenfranchised because they have no power of interaction, is real. In the hands of the unscrupulous such power could be mis-applied - and there is no legal constraint presently available to any regulator to prevent such misuse as these services are one-to-one and can be different for every viewer. There is no provision in the 1990 Broadcasting Act (or the earlier 1984 Cable and Broadcasting Act) to cater for this situation. The only regulation is the common law, similar to that which controls the printed press - yet the letters column and real-time interactivity are miles apart in their impact. Expressing an opinion about political matters is a fundamental right which is regulated only by privilege. Why should technology means of exercising that right be any different from traditional paper ?

Only time will tell whether there is an absence of any effective legal instruments. Such absence would be in line with current Government 'market driven' policies but the technical success or failure of new services could be determined by conscious investment, rather than demand. The possibility is that intervention in the case of a social or political misuse of interactive mechanisms may be virtually impossible under current legislation, and any steps to correct that position could take months, if not years.

This last idea may seem extreme, and may not be 'very English'. However we should consider that much of our media infrastructure is not UK owned and there is no right of access to the huge amounts of channel space and programme airtime which are becoming available.

Despite the energy and the preparedness to experiment with media which has long been evidenced in the British voluntary sector, the inherent conservatism of multi-national infrastructure companies means that it is likely that the energy in developing interactive services will come from national and international sources and the local opportunity may be ignored. This may be a suitable recipe for profit-taking but could it also lead to exploitation in the hands of companies (or their controlling interests) who are becoming used to operating beyond the reach of a single nation's law.

The future of local television, interactive or not, is now becoming bound up in the right of access to distribution networks.

Cable companies have mostly, so far, reneged on promises expressed in their franchise applications and Parliament has given no effective powers to the regulators to ensure that local TV services can be created even when the networks are fully built. A cable company is the absolute gateway to its own network and any refusal to implement local TV, or to allow others to act in creating local services, is a court without appeal.

Jon Davey at the ITC is ready to point out that UK regulators or legislators cannot force companies to spend money when those companies deem such expenditure to be against their commercial interests. But he is equally clear that this stance applies to building the networks, not to the services that are carried on them. He believes that operators should be prepared to spend money (as we have seen the sums involved are relatively small) on local services which do not necessarily bring them a direct return but which should benefit them in other ways.

Consideration should now be given to regulation which requires local networks to carry local services where such services are offered by local third parties. A standard fee structure could be created and a tougher licensing regime introduced to ensure that time-wasters can be weeded out early on. Only in this way can the rapid changes that are now taking place in technology develop into services which reflect the needs of local communities. Only in this way can development of services be driven by the market !

Such developments would benefit cable companies. They would gain desirable services tuned to the needs of their audience, and of niche groups in the local community, which should add to their subscriber base - whether for the TV or the telephone.

Whatever the failings of the last twelve years the 1982 ITAP report conjured visions which remain valid and technically appropriate today. Local television could also take on new dimensions not dreamed of in 1982. If it continues to be stifled at the local level this important communications opportunity will eventually become political, to the ultimate detriment of the cable companies.

APPENDIX

Some Useful Addresses

REGULATORY AND REPRESENTATIVE BODIES

Independent Television Commission
33 Foley Street
London
W1P 7LB
Director of Cable and Satellite: Jon Davey
Senior Programme Officer: Guy Phelps
Cable Development Officer: Carmel McLaughlin
The television regulatory and licensing body. The Cable and Satellite Division is responsible for franchising cable systems, licensing programme channels and the associated regulatory oversight.

Cable Television Association
Artillery House
Artillery Row
London
SW1P 1RT
Director General: Richard Woollam
Representative organisation for cable operators, their owners and associates, and for leading suppliers to the industry. Organises the annual European Cable Convention, usually held in October.

The Guild of Local Television
16 Fountain Road
Birmingham
B17 8NL
Independent organisation dedicated to furthering the cause of local television and to supporting practitioners.

Anyone seeking the addresses and contacts of individual cable operating companies is recommended to contact the CTA for the information in the first instance. The addresses and contacts of individual licensed TV channels can be obtained from the ITC. The

ITC publishes an annual Factfile which contains the contact details of all its licensees. This is available from the Information Office at the above address.

JOURNALS AND PUBLICATIONS

Inside Cable
PO Box 5
Church Stretton
Shropshire
SY6 6ZZ
Editor: Roger Wilson
Low cost newsletter covering the UK cable industry and its range of services, including local television.

Cable and Satellite Europe
21st Century Publications
531-533 Kings Road
London
SW10 0TZ
Monthly magazine covering the cable and satellite industry in Europe and reporting news from around the world.

CSC - Cable and Satellite Communications International
104 City View
463 Bethnal Green Road
London
E2 9QY
Monthly magazine with a focus on UK cable.

Who's Who in Cable and Satellite
WOAC Communications
Claremont House
Whipsnade
Dunstable
LU6 2LH
Editor: Maureen Morgan
The UK's industry directory published in January and July each year. WOAC also publishes The Cable TV and Telecom Yearbook - an authoritative annual review of the industry.